SpringerBriefs in Applied Sciences and Technology

Series editor

Janusz Kacprzyk, Polish Academy of Sciences, Systems Research Institute, Warsaw, Poland

SpringerBriefs present concise summaries of cutting-edge research and practical applications across a wide spectrum of fields. Featuring compact volumes of 50 to 125 pages, the series covers a range of content from professional to academic. Typical publications can be:

- A timely report of state-of-the art methods
- An introduction to or a manual for the application of mathematical or computer techniques
- A bridge between new research results, as published in journal articles
- A snapshot of a hot or emerging topic
- An in-depth case study
- A presentation of core concepts that students must understand in order to make independent contributions

SpringerBriefs are characterized by fast, global electronic dissemination, standard publishing contracts, standardized manuscript preparation and formatting guidelines, and expedited production schedules.

On the one hand, **SpringerBriefs in Applied Sciences and Technology** are devoted to the publication of fundamentals and applications within the different classical engineering disciplines as well as in interdisciplinary fields that recently emerged between these areas. On the other hand, as the boundary separating fundamental research and applied technology is more and more dissolving, this series is particularly open to trans-disciplinary topics between fundamental science and engineering.

Indexed by EI-Compendex and Springerlink.

More information about this series at http://www.springer.com/series/8884

Lakshminarayan Hazra · Pubali Mukherjee

Self-similarity in Walsh Functions and in the Farfield Diffraction Patterns of Radial Walsh Filters

 Springer

Lakshminarayan Hazra
Department of Applied Optics
and Photonics
University of Calcutta
Kolkata, West Bengal
India

Pubali Mukherjee
Department of Electronics
and Communication Engineering
MCKV Institute of Engineering
Howrah, West Bengal
India

ISSN 2191-530X   ISSN 2191-5318   (electronic)
SpringerBriefs in Applied Sciences and Technology
ISBN 978-981-10-2808-3   ISBN 978-981-10-2809-0   (eBook)
DOI 10.1007/978-981-10-2809-0

Library of Congress Control Number: 2017935390

Printed on acid-free paper

This Springer imprint is published by Springer Nature
The registered company is Springer Nature Singapore Pte Ltd.
The registered company address is: 152 Beach Road, #21-01/04 Gateway East, Singapore 189721, Singapore

Preface

Walsh functions form a closed set of orthogonal functions over a prespecified interval, each function taking merely one constant value (either +1 or −1) in each of a finite number of subintervals into which the entire interval is divided. Order of a Walsh function is equal to the number of zero crossing within the interval. Walsh functions are extensively used in communication theory and microwave engineering, and in the field of digital signal processing. One of the authors (LNH) pioneered the use of Walsh functions in the treatment of problems of optical imagery and beam shaping. Lossless phase filters play the key role in these ventures. Walsh filters, derived from Walsh functions, have opened up new vistas. They take on values, either 0 or π phase, corresponding to +1 or −1 value of the Walsh function.

In general, members of the set of Walsh functions are not self-similar; but the set can be classified into distinct self-similar groups and subgroups, where members of each subgroup possess distinct self-similar structures. Self-similarity is observed in the farfield diffraction patterns obtained on the focal/image plane of optical imaging systems when the corresponding self-similar filters are used for pupil plane filtering. Self-similarity is also observed in the three-dimensional (3D) light distributions around the focus obtained by use of self-similar pupil plane Walsh filters. Our observations on the same are put forward in the monograph. They provide valuable clues in tackling the inverse problem of synthesis of phase filters for prespecified distribution of intensity on the transverse image/focal plane, or general problem of prespecified 3D distribution of intensity around the image/focal plane.

Authors wish to acknowledge support and encouragement provided by their spouses, namely Sukla Hazra and Shankha Bose, in preparation of this monograph.

Kolkata, India   Lakshminarayan Hazra
April 2016   Pubali Mukherjee

Contents

About the Authors

Prof. Lakshminarayan Hazra is a B.Sc. (Hons) physics graduate and postgraduate of applied physics with a doctorate from the University of Calcutta, Kolkata. Prof. Lakshminarayan Hazra has over four decades of academic and industrial experience. He is an Emeritus Professor and Former Head of the Department of Applied Optics and Photonics at the University of Calcutta, Kolkata, India. His areas of professional specialization include lens design/optical system design, image formation and aberration theory, diffractive optics, and optical and photonic instrumentation. He is a Fellow of the Optical Society of America, and the International Society for Optics and Photonics (SPIE). He is the editor in chief of the archival journal, Journal of Optics, published by M/s Springer in collaboration with the Optical Society of India. He has published more than 150 journal articles and books.

Pubali Mukherjee holds B.Sc. (Hons.), B.Tech., M.Tech., and Ph.D. degrees, all from the University of Calcutta. Currently, she is an assistant professor in Electronics and Communication Engineering Department at the MCKV Institute of Engineering, Howrah, West Bengal, India. She has 10 years of teaching and 5 years of research experience. Her areas of interest include optical systems, image assessment criteria, and diffraction pattern tailoring using phase filters and applications. She has published many papers in journals and conference proceedings.

Chapter 1
Walsh Functions

Abstract With a brief description of the origin, basic characteristics and major practical applications of Walsh functions, different forms of Walsh functions in one and two dimensions are systematically developed from the orthogonality considerations. Approximation of a continuous function over a given domain by a finite set of Walsh functions provides a piecewise constant approximation of the function. The latter has important consequences in tackling practical problems by way of opening up new techniques for analysis and synthesis. Highlights of these aspects are provided along with a brief description of Walsh Block functions and Hadamard matrices.

Keywords Walsh functions · Polar Walsh functions · Radial Walsh functions · Azimuthal Walsh functions · Annular Walsh functions

1.1 Introduction

Walsh functions were introduced by J.L. Walsh in 1923 as 'a closed set of normal orthogonal functions' defined over the interval (0, 1) and having values $+1$ or -1 within the interval [1]. They are a complete set of normal orthogonal functions over a given finite interval and take on values $+1$ or -1, except at a finite number of points of discontinuity, where they take the value 0. The order of the Walsh function is equal to the number of zero crossings or phase transitions within the specified domain over which the set is defined. The complete set of Walsh functions not only provides a viable alternative but also introduces a significant simplification in the approach of using suitable base functions [2]. They have the interesting property that an approximation of a continuous function over a finite interval by a finite number of base functions of this set leads to a piecewise-constant approximation to the function. But until recently these functions have not been used to any significant extent in the treatment of physical problems [3]. However, it is now

© The Author(s) 2018

L. Hazra and P. Mukherjee, *Self-similarity in Walsh Functions and in the Farfield Diffraction Patterns of Radial Walsh Filters*, SpringerBriefs in Applied Sciences and Technology, DOI 10.1007/978-981-10-2809-0_1

being realized that the system of Walsh functions possesses some unique properties among all the systems orthonormal in Hilbert space L_2 (0, 1). In fact, the system of Walsh functions may be derived from the character group of the dyadic group which is a topologic group derived from the set of binary representations of the real numbers. This topology of the dyadic group makes properties of the system of Walsh functions strikingly different from those of other systems [4]. Walsh functions may be defined for a single variable or for two variables in rectangular coordinates [5]. Formulation in polar coordinates can be made via the orthogonality properties [2, 6]. Radial Walsh functions and annular Walsh functions can be derived from the polar Walsh functions [7, 8]. Annular Walsh functions are a generalization of radial Walsh functions. They are specific to a particular value of central obscuration ε. Annular Walsh functions which form a complete set of orthogonal functions over the annular unobscured circle take on values either $+1$ or -1 over different annular subdomains specified by the inner and outer radii of the annulus. The value of any annular Walsh function is taken as zero from the centre of the circular aperture to the inner radius of the unobscured annulus. Walsh functions have been utilized to a great advantage in the field of signal coding and transmission, in allied problems of information processing [3, 4, 8] and in digital signal processing [4]. Two dimensional Walsh functions in the usual rectangular co-ordinates have been extensively used in digital image processing applications [5]. For the treatment of problems of image formation where one usually deals with some form of axial symmetry, it is useful to develop Walsh functions in polar co-ordinates. For systems with rotational symmetry about the axis, it is sufficient to deal only with radial Walsh functions. The latter have been developed as a special case of Walsh functions in polar co-ordinates [2] and have been found to be quite useful in the treatment of problems of apodization [9–11] and the studies on aberrated optical imagery.

1.2 One Dimensional Walsh Functions

A Walsh function of a single variable x is represented as $W_k(x)$ where the subscript k refers to the order of the function. The function takes on values either $+1$ or -1 over a specified domain and the set of Walsh functions is orthogonal and complete in the interval. The same implies the following relation for the set of Walsh functions defined over the interval (0, 1).

$$\int_0^1 W_k(x)W_\ell(x)dx = \delta_{k\ell} \tag{1.1}$$

where $\delta_{k\ell}$ is the Kronecker delta defined as

$$\delta_{k\pounds} = \begin{cases} 0, & k \neq \pounds \\ 1, & k = \pounds \end{cases} \tag{1.2}$$

The set has a further specification through the unique definition of the zeroth element of the set as

$$W_0(x) = 1, \quad 0 \leq x \leq 1 \tag{1.3}$$

Figure 1.1 show the first four one-dimensional Walsh functions $W_k(x)$, $k = 0$, 1, 2, 3.

In order to obtain an analytical representation for the unidimensional Walsh function $W_k(x)$, let the integer k be expressed as

$$k = \sum_{i=0}^{v-1} k_i 2^i \tag{1.4}$$

where k_i are the bits, 0 or 1 of the binary numeral for k and 2^v is the integral power of 2 that just exceeds k. Thus for $k = 0$, 1, $v = 1$; for $k = 0$, 1, 2, 3, $v = 2$; for $k = 0$, 1, 2, 3, 4, 5, 6, 7, $v = 3$, and so on.

For all x in the interval $(0, 1)$ we define

$$W_k(x) = \prod_{i=0}^{v-1} \mathrm{sgn}\left[\cos\left(2^i \pi x k_i\right)\right] \tag{1.5}$$

Fig. 1.1 One dimensional Walsh functions, $W_k(x)$, $k = 0, 1, \ldots, 3$

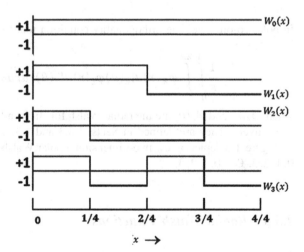

where

$$\begin{aligned}
\text{sgn}(x) \quad &= +1 \quad x > 0 \\
&= 0 \quad\quad x = 0 \\
&= -1 \quad\; x < 0
\end{aligned} \tag{1.6}$$

It may be noted from Fig. 1.1 that the order of the Walsh function is given by the number of zero crossings, i.e. the sign changes of the function in the interval $(0, 1)$.

1.3   Two Dimensional Walsh Functions

1.3.1   Rectangular Walsh Functions

In two dimensions, the orthogonality condition takes the form

$$\int_0^1 \int_0^1 W_{\hbar}(x) W_{\epsilon}(y) W_{\pounds}(x) W_{¥}(y) dx dy = \delta_{\hbar\pounds}\delta_{\epsilon¥} \tag{1.7}$$

Figure 1.2 depicts a few two dimensional rectangular Walsh functions $W_{\hbar}(x) W_{\epsilon}(y)$, $\hbar = 0, 1, 2, 3$, $\epsilon = 0, 1, 2, 3$.

1.3.2   Polar Walsh Functions

In polar co-ordinates the orthogonality condition reduces to

$$\frac{1}{\pi} \int_0^1 \int_0^{2\pi} \varphi_{\hbar}(r) \mathcal{A}_{\epsilon}(\theta) \varphi_{\pounds}(r) \mathcal{A}_{¥}(\theta) r dr d\theta = \delta_{\hbar\pounds}\delta_{\epsilon¥} \tag{1.8}$$

where $\varphi_{\hbar}(r)$ and $\mathcal{A}_{\epsilon}(\theta)$ are the radial Walsh function and azimuthal Walsh function respectively. They are defined in Sects. 1.3.3 and 1.3.4.

Figure 1.3 depicts a few two-dimensional polar Walsh functions $\varphi_{\hbar}(r) \mathcal{A}_{\pounds}(\theta) \hbar = 0, 1, 2, 3$, $\epsilon = 0, 1, 2, 3$.

1.3.3   Radial Walsh Functions

Alike the case of unidimensional Walsh functions, $W_{\hbar}(x)$, radial Walsh functions [2] $\varphi_{\hbar}(r)$ of index $\hbar \geq 0$ and argument r over the interval $(0, 1)$ require integer $\hbar$ to be expressed in the form.

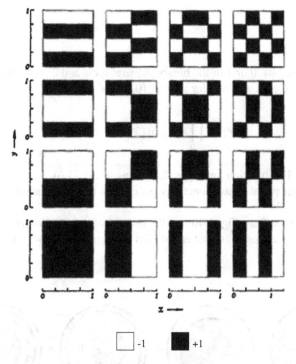

$\square$ -1 $\blacksquare$ +1

Fig. 1.2 Two dimensional rectangular Walsh functions $W_k(x)W_\ell(y)$, $k = 0, 1, 2, 3$, $\ell = 0, 1, 2, 3$

Fig. 1.3 Two dimensional polar Walsh functions $\varphi_k(r)\mathcal{A}_\ell(\theta)$, $k = 0, 1, 2, 3$ $\ell = 0, 1, 2, 3$ radial Walsh functions

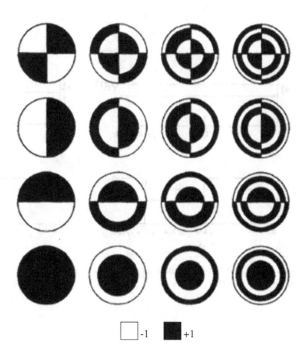

$\square$ -1 $\blacksquare$ +1

$$k = \sum_{i=0}^{\gamma-1} k_i 2^i \qquad (1.9)$$

where k_i are the bits, 0 or 1 of the binary numeral for k and 2^γ is the integral power of 2 that just exceeds k. Thus for all r in $(0,1)$ we define

$$\varphi_k(r) = \prod_{i=0}^{\gamma-1} \mathrm{sgn}\left[\cos\left(2^i \pi t k_i\right)\right] = \varphi_k(t) \qquad (1.10)$$

where $t = r^2$. The function $\mathrm{sgn}(x)$ is defined as in Eq. (1.6).

The first four radial Walsh functions $\varphi_k(r)$, $k = 0, 1, 2, 3$ are shown in. (r, θ). space in Fig. 1.4b.

It may be noted from Fig. 1.4b that the order k of the function $\varphi_k(r)$ is equal to the number of zero crossings, that is, the sign changes of the function in the interval $(0, 1)$.

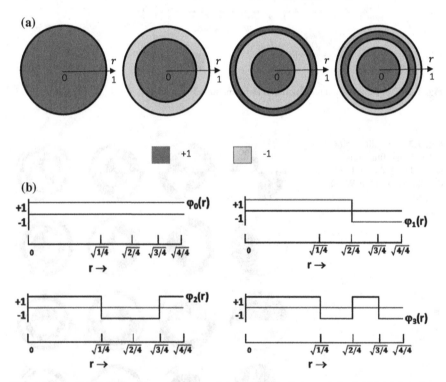

Fig. 1.4 a Radial Walsh functions $\varphi_k(r)$, $k = 0, 1, \ldots, 3$ in (r, θ) space. **b** Radial Walsh functions $\varphi_k(r) - r$, $k = 0, 1, \ldots, 3$.

For azimuth invariant case the orthogonality condition takes the form

$$\int_0^1 \varphi_\ell(r)\varphi_\pounds(r)rdr = \frac{1}{2}\delta_{\ell\pounds} \tag{1.11}$$

where $\delta_{\ell\pounds}$ is the Kronecker delta defined as in Eq. (1.2).

In $t\,(=r^2)$ space radial Walsh functions $\varphi_\ell(t)$ satisfy the orthogonally condition

$$\int_0^1 \varphi_\ell(t)\varphi_\pounds(t)dt = \delta_{\ell\pounds} \tag{1.12}$$

The radial Walsh functions $\varphi_\ell(t)$, $\ell = 0, 1, \ldots, 3$ are shown in Fig. 1.5.

1.3.4   Azimuthal Walsh Functions

In order to obtain analytical expression for azimuthal Walsh functions of $\mathcal{A}_\ell(\theta)$ of index $\ell \geq 0$ and argument θ, is required to express the integral index ℓ the form

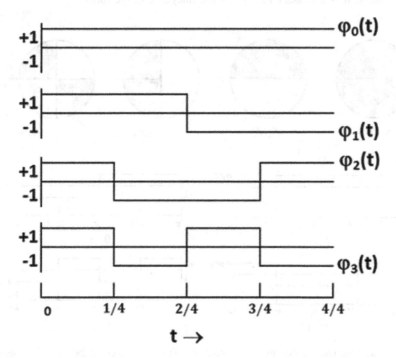

Fig. 1.5 Radial Walsh functions in t space, $\varphi_\ell(t), \ell = 0, 1, \ldots, 3$

$$\text{€} = \sum_{i=0}^{v-1} \text{€}_i 2^i \tag{1.13}$$

where €_i are the bits 0 or 1 of the binary numeral for €, and 2^v is the integral power of 2 that just exceeds €. For all θ in the interval $(0, 2\pi)$ we define the azimuthal Walsh function $\mathcal{A}_{\text{€}}(\theta)$ as

$$\mathcal{A}_{\text{€}}(\theta) = \prod_{i=0}^{v-1} \text{sgn}\left[\cos\left(2^i \frac{\theta}{2} \text{€}_i\right)\right] \tag{1.14}$$

where $\text{sgn}[x]$ is defined in Eq. 1.6.

The first four azimuthal Walsh functions $\mathcal{A}_{\text{€}}(\theta), \theta = 0, 1, 2, 3$ are shown in Fig. 1.6.

1.3.5   Annular Walsh Functions

Alike the earlier cases, to define annular Walsh function [7, 12] $\varphi_{\ell}^{\varepsilon}(r)$ of index $\ell \geq 0$ and argument r over an annular region with ε and 1 as inner and outer radii respectively, it is necessary to express the integer ℓ in the form

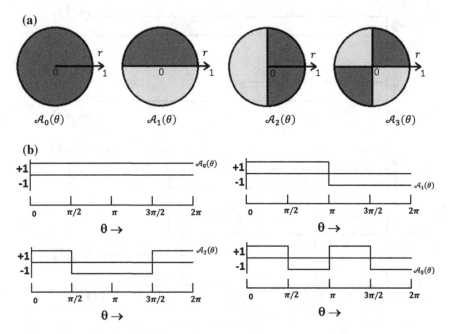

Fig. 1.6 a Azimuthal Walsh functions $\mathcal{A}_{\text{€}}(\theta), \text{€} = 0, 1, \ldots, 3$ over $(0, 2\pi)$ in (r, θ) space. **b** Azimuthal Walsh functions $\mathcal{A}_{\text{€}}(\theta) - \theta, \text{€} = 0, 1, \ldots, 3$. over $(0, 2\pi)$

$$k = \sum_{m=0}^{\gamma-1} k_m 2^m \tag{1.15}$$

k_m are the bits, 0 or 1 of the binary numeral for k, and (2^γ) is the power of 2 that just exceeds k. For all r in $(\varepsilon, 1)$, $\varphi_k^\varepsilon(r)$ is defined as

$$\varphi_k^\varepsilon(r) = \prod_{m=0}^{n-1} \text{sgn}\left\{\cos\left[k_m 2^m \pi \frac{(r^2 - \varepsilon^2)}{(1 - \varepsilon^2)}\right]\right\} \tag{1.16}$$

where $\text{sgn}(x)$ is defined as in Eq. (1.6).

The orthogonality condition implies

$$\int_\varepsilon^1 \varphi_k^\varepsilon(r)\varphi_\mathcal{L}^\varepsilon(r)rdr = \frac{1 - \varepsilon^2}{2}\delta_{k\mathcal{L}} \tag{1.17}$$

where $\delta_{k\mathcal{L}}$ is the Kronecker delta defined as in Eq. (1.2)

Figure 1.7 shows the first four annular Walsh functions for central obscuration ratio $\varepsilon = 0.3$. Figure 1.8 presents values of the functions $\varphi_k^\varepsilon(r), k = 0, \ldots, 3$ along the radius in an azimuthal direction. It should be noted that the order of the functions, n is equal to the number of zero crossings, or sign changes of the function in the interval $(0.3, 1)$, and locations of the points of zero crossings for members of the set of functions $\varphi_k^{0.3}(r), k = 0, \ldots, 3$ are given by

$$r_i = \sqrt{\frac{[(4 - i)\varepsilon^2 + i]}{4}}, \quad i = 1, 2, 3 \tag{1.18}$$

In general, for the first, where $N = 2^b$, and b is a positive integer, annular Walsh functions $\varphi_k^\varepsilon(r), k = 0, 1, \ldots, (N - 1)$, the zero crossings are located at

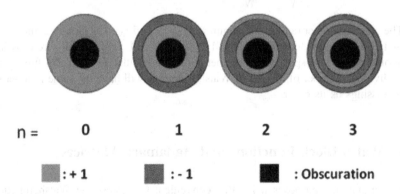

n =   0   1   2   3

:+1   :-1   : Obscuration

Fig. 1.7 Annular Walsh functions. $\varphi_k^\varepsilon(r)$ of order $k = 0, 1, 2, 3$ in (r, θ) space for central obscuration ratio $\varepsilon = 0.3$.

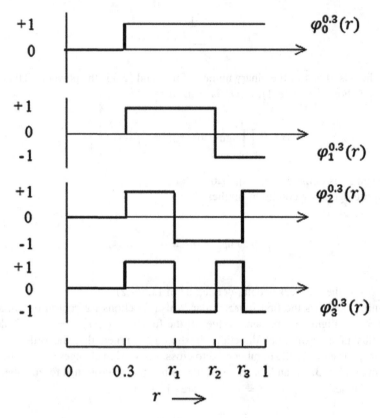

Fig. 1.8 Annular Walsh functions $\varphi_k^\varepsilon(r)$ of order $k = 0, 1, 2, 3$. along radius r for central obscuration ratio $\varepsilon = 0.3$.

$$r_i = \sqrt{\frac{[(N-i)\varepsilon^2 + i]}{N}}, \quad i = 1, 2, \ldots, (N-1) \tag{1.19}$$

The inner and outer radii of the annulus is $r_0 = \varepsilon$ and $r_N = 1$. Note that the set of $(N-1)$ zero crossing locations, $r_i, i = 1, 2, \ldots, (N-1)$ consist of all zero crossing locations required for specifying members of this particular set of Walsh functions. An individual member of this set of Walsh functions will have the same number of zero crossings as its order.

1.4   Walsh Block Functions and Hadamard Matrices

For computational purposes it is often convenient to express the Walsh function $W_k(x)$ as

$$W_k(x) = \sum_{j=1}^{N} H_{kj} B_j^N(x) \tag{1.20}$$

where $N = 2^\gamma$. For a particular value of k, the integer γ is taken such that N just exceeds k. $B_j^N(x)$ are zero-one functions, also known as Walsh Block functions, defined as

$$B_j^N(x) = \begin{cases} 1, & x_{j-1} \leq x \leq x_j \\ 0, & \text{otherwise} \end{cases} \tag{1.21}$$

For all Walsh functions of order $k < N$, the same values of x_j, $j = 0, N$ are to be used. They are unique for the particular value of N, and their values are given by

$$x_j = \left(\frac{j}{N}\right), \quad j = 1, 2, 3, \ldots (N-1) \tag{1.22}$$

H_{kj} are the elements of a $(N \times N)$ Hadamard matrix whose elements are $+1$ or -1 [3]. Hadamard matrices for any value of N can be obtained from the defining relation (1.5) of Walsh functions (Fig. 1.9).

Hadamard matrices for $N = 2^2$ and $N = 2^3$ are given in Fig. 1.10a, b respectively.

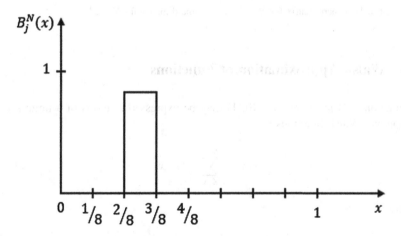

Fig. 1.9 Walsh Block function $B_j^N(x)$ for $N = 8$ and $j = 3$

(a)

k \ j		1	2	3
0	+1	+1	+1	+1
1	+1	+1	-1	-1
2	+1	-1	-1	+1
3	+1	-1	+1	-1

(b)

k \ j	0	1	2	3	4	5	6	7
0	+1	+1	+1	+1	+1	+1	+1	+1
1	+1	+1	+1	+1	-1	-1	-1	-1
2	+1	+1	-1	-1	-1	-1	+1	+1
3	+1	+1	-1	-1	+1	+1	-1	-1
4	+1	-1	-1	+1	+1	-1	-1	+1
5	+1	-1	-1	+1	-1	+1	+1	-1
6	+1	-1	+1	-1	-1	+1	-1	+1
7	+1	-1	+1	-1	+1	-1	+1	-1

Fig. 1.10 a Hadamard matrix for $N = 2^2$. **b** Hadamard matrix for $N = 2^3$

1.5   Walsh Approximation of Functions

A function $f(x)$ in the domain $(0, 1)$ may be expressed in terms of a finite set of orthogonal Walsh functions as

$$f(x) = \sum_{k=0}^{N-1} a_k W_k(x) \tag{1.23}$$

where

$$a_h = \int_0^1 f(x) W_k(x) dx \tag{1.24}$$

Fig. 1.11 A $4 = (2^2)$ step ladder approximation of $f(x)$ by using first four Walsh functions. The fixed value in each step is equal to the average value of $f(x)$ over the subinterval

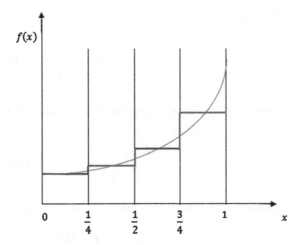

A finite sum of N terms will lead to a ladder approximation with P steps (Fig. 1.11). An interesting property of Walsh approximation is that the number of steps P is given by

$$P = 2^p \tag{1.25}$$

For all N in the interval

$$2^{p-1} < N \le 2^p, \quad p = 1, 2, \ldots \tag{1.26}$$

Walsh approximation of functions provide highly useful relations when the number of terms in approximation is taken as $N = P = 2^p$, for the values of the function in each ladder step then becomes exactly equal to the average value of the function over the subdomain.

The value of the function at the jth step is given by

$$f_j = \sum_{k=0}^{N-1} a_k H_{kj}, \quad j = 1, \ldots, P \tag{1.27}$$

where H_{kj} are elements of a Hadamard matrix defined earlier. The values of these elements are either $+1$ or -1, and is determined by the order of the Walsh function k and the number of the step, j.

Following the definition of Walsh Block functions given in the earlier section, $W_k(x)$ can be represented as

$$W_k(x) = \sum_{j=1}^{N} H_{kj} B_j^P(x) \tag{1.28}$$

From (1.24) it follows

$$a_k = \int_0^1 f(x)W_k(x)dx = \sum_{j=1}^{P} H_{kj} \int_0^1 f(x)B_j^P(x)dx \qquad (1.29)$$

Using the property of Walsh Block functions one gets

$$a_k = \sum_{j=1}^{P} H_{kj} \int_{x_{j-1}}^{x_j} f(x)dx = \left(\frac{1}{P}\right) \sum_{j=1}^{P} H_{kj}\bar{f}_j \qquad (1.30)$$

In matrix notation the relations between (1.27) and (1.30) can be written as

$$\boldsymbol{f} = \boldsymbol{Ha} \qquad (1.31)$$

$$\boldsymbol{a} = \left(\frac{1}{P}\right)\boldsymbol{H\bar{f}} \qquad (1.32)$$

where $\boldsymbol{H}$ is the symmetric Hadamard matrix, a is the column vector of the coefficients a_k, f and $\bar{f}$ are the column vectors of the values of the function at the ladder steps f_j as given by the Walsh approximation, and the average value of the function over the subdomain of the step respectively. From (1.31) to (1.32) it follows

$$\boldsymbol{f} = \left(\frac{1}{P}\right)\boldsymbol{HH\bar{f}} = \left(\frac{1}{P}\right)\boldsymbol{PI\bar{f}} = \boldsymbol{I\bar{f}} \qquad (1.33)$$

where $\boldsymbol{I}$ is the identity matrix. Therefore, it follows that the values of the function at each ladder step of Walsh approximation of the function with finite number of terms $N = P = 2^p$, with integral values for p, is equal to the average value of the continuous function $f(x)$ over the subdomain of the step. An illustration is given in Fig. 1.11.

1.6   Multiplication Rule for Walsh Functions

The product of two Walsh functions over a domain is a third Walsh function over the same domain given according to the rule

$$W_m(x)W_n(x) = W_{m\oplus n}(x) \qquad (1.34)$$

where the operator $\oplus$ notes binary addition, i.e. addition modulo 2, without carry.

References

1. Walsh JL (1923) A closed set of normal orthogonal functions. Am J Math 45(1):5–24
2. Hazra LN, Banerjee A (1976) Application of Walsh function in generation of optimum apodizers. J Opt (India) 5:19–26
3. Harmuth HF (1972) Transmission of information by orthogonal functions. Springer, Berlin, p 31
4. Beauchamp KG (1985) Walsh functions and their Applications. Academic Press, New York
5. Andrews HC (1970) Computer techniques in image processing. Academic, New York
6. De M, Hazra LN (1977) Walsh functions in problems of optical imagery. Opt Acta 24 (3):221–234
7. Hazra LN (2007) Walsh filters in tailoring of resolution in microscopic imaging. Micron 38 (2):129–135
8. Hazra LN, Guha A (1986) Farfield diffraction properties of radial Walsh filters. J Opt Soc Am A 3(6):843–846
9. Hazra LN, Purkait PK, De M (1979) Apodization of aberrated pupils. Can J Phys 57(9):1340–1346
10. Purkait PK (1983) Application of Walsh functions in problems of aberrated optical imagery, Ph.D. Dissertation, University of Calcutta, India
11. Hazra LN (1977) A new class of optimum amplitude filters. Opt Commun 21(2):232–236
12. Mukherjee P, Hazra, LN (2013) Farfield diffraction properties of annular Walsh filters. Adv Opt Tech 2013, ID 360450, 6 pages,

Chapter 2
Self-similarity in Walsh Functions

Abstract A structure which can be divided into smaller and smaller pieces, each piece being an exact replica of the entire structure is called self-similar. The set of Walsh functions can be classified into distinct self-similar groups and subgroups where members of each subgroup exhibit self-similarity. After a brief discussion on the generation of higher order Walsh functions from lower order Walsh functions by an alternating process, a scheme for classification of Walsh functions into self-similar groups and subgroups is presented. Self-similarity in radial and annular Walsh functions and the correspondence between Walsh filters and Walsh functions are also discussed.

Keywords Self-similarity · Fractals · Fractal optics · Self-similar Walsh filters

2.1 Introduction

Self-similarity is a typical property of fractals. The fractal structures play an important role in describing and understanding a large number of phenomena in several areas of science and technology [1]. In optics, the fractal structure of some optical wave fields and the diffraction patterns generated from various fractal apertures are examples of this trend [2, 3]. A family of shapes that are irregular and fragmented and cannot be explained by standard geometry can be put under a class called fractals, a term coined by Mandelbrot in 1975. French-American polymath Benoit Mandelbrot (1924–2010) is often referred to as the 'Father of Fractals'. In his 1982 book, 'The Fractal Geometry of Nature' [4], he noted that the word fractal comes from the Latin word 'fractus', meaning broken. A structure may be called fractal if it possesses some or all of the following properties namely, self-similarity, fine-structure, (that is, it preserves details at all scales however small), size of the structure depends on the scale at which it is measured, classical methods of geometry and mathematics are not applicable to the structure, the structure has a natural appearance and the structure has a simple recursive construction.

© The Author(s) 2018

L. Hazra and P. Mukherjee, *Self-similarity in Walsh Functions and in the Farfield Diffraction Patterns of Radial Walsh Filters*, SpringerBriefs in Applied Sciences and Technology, DOI 10.1007/978-981-10-2809-0_2

A structure is said to be self-similar when it can be divided into smaller and smaller pieces, each piece being an exact replica of the entire structure [5]. The Von Koch curve, Sierpinski triangle, Cantor fractals are some of the very simple and common examples of fractals. Fractals with completely different appearance may be constructed by a very similar recursive procedure. The Von Koch curve was obtained by taking a straight line, dividing it into three equal parts, erasing the middle part and replacing it by the other sides of an equilateral triangle on the same base. This gives a chain of four shorter joined up straight line segments and the curve is finally obtained by doing the same thing to each of the four pieces, that is, by repeatedly replacing each straight line segment by a simple figure ⌐ᴧ⌐ sometimes called a generator or motif of the curve. The curve is self-similar, that is, it is made up of smaller copes of itself. In particular we could cut it into 4 parts each a 1/3 scale copy of the entire curve. The Cantor fractal is perhaps the most basic self-similar fractal called the middle third Cantor set. Each stage of this construction is obtained by removing the middle third of each interval in the preceding stage, that is, it is made up of 2 scale 1/3 copies of itself.

The complete set of Walsh functions do not exhibit self-similarity among their individual constituents. But members of the set can be classified into distinct groups, individual members of each group exhibiting self-similarity among themselves. In this chapter we intend to explore the inherent self-similarity present in the various orders of the Walsh functions and classify them into self-similar groups and find analytical expressions to represent members of such self-similar groups. Our study is mostly limited to orders ℓ within the range (0, 15) with occasional foray in Walsh functions of neighbouring higher orders. Discussion on the same is put forward in subsequent sections.

2.2 Generation of Higher Order Walsh Functions from Lower Order Walsh Functions

Unidimensional Walsh functions, $W_\ell(x)$, for order $\ell = 0, 1, \ldots, 15$ are presented succinctly in Fig. 2.1. The total domain for each of the Walsh functions is (0, 1). For every Walsh function, the total domain consists of a finite number of equal subdomains. The number of subdomains depends on the order of the Walsh function. The minimum number, say P, of subdomains to represent a Walsh function is an integral power of 2. For representing $W_0(x), P = 2^0 = 1$. For other orders of Walsh functions P is given by the relation $P = 2^{\tilde{m}}$, where $\tilde{m}$, a positive integer, satisfies the relation $2^{\tilde{m}-1} < \ell \leq 2^{\tilde{m}}$. For example, the number of equal parts, P, is 2 for $W_1(x)$, is 4 for each of $W_2(x)$ and $W_3(x)$, and 8 for each of the Walsh functions $W_4(x), W_5(x), W_6(x)$ and $W_7(x)$ Again, P becomes 16 for each of the Walsh functions $W_i(x), i = 8, \ldots, 15$.

In Fig. 2.1 each row consists of elements comprised of either '+' or '−' denoting values +1 or −1 respectively. The first row represents $W_0(x)$ that consists of a single

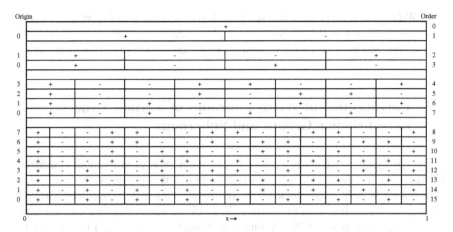

Fig. 2.1 Each row represents unidimensional Walsh functions $\mathbf{W}_\ell(x)$ over the interval (0, 1) for $\ell = 0, 1, \ldots, 15$ in ascending order of sequence

domain (0, 1) with '+' over it. The second row represents $W_1(x)$. It consists of two subdomains; the first subdomain extending over $\left(0, \frac{1}{2}\right)$ has '+' over it, and the second subdomain extending over $\left(\frac{1}{2}, 1\right)$ has '−' over it. Each of $W_2(x)$ and $W_3(x)$ consists of four subdomains $\left(0, \frac{1}{4}\right)$, $\left(\frac{1}{4}, \frac{1}{2}\right)$, $\left(\frac{1}{2}, \frac{3}{4}\right)$ and $\left(\frac{3}{4}, 1\right)$, and they are shown in the third and the fourth row respectively. Each of the four Walsh functions $W_i(x), i = 4, \ldots, 7$ consists of eight subdomains $\left(0, \frac{1}{8}\right)$, $\left(\frac{1}{8}, \frac{2}{8}\right)$, $\left(\frac{2}{8}, \frac{3}{8}\right)$, $\left(\frac{3}{8}, \frac{4}{8}\right)$, $\left(\frac{4}{8}, \frac{5}{8}\right)$, $\left(\frac{5}{8}, \frac{6}{8}\right)$, $\left(\frac{6}{8}, \frac{7}{8}\right)$ and $\left(\frac{7}{8}, 1\right)$, and they are shown in the corresponding $(i + 1)$th rows in Fig. 2.1. Walsh functions $W_i(x), i = 8, \ldots, 15$ are presented similarly in corresponding $(i + 1)$th row. Each of them consists of 16 equal subdomains $\left(0, \frac{1}{16}\right)$, $\ldots$, $\left(\frac{15}{16}, 1\right)$.

To obtain Walsh functions of higher order from Walsh functions of lower order an 'alternating process' can be adopted [6]. The number on the extreme right of each row in Fig. 2.1 gives the order of the Walsh function corresponding to that row, and the number on the extreme left of each row gives the order of the Walsh function from which it is derived by the alternating process. $W_1(x)$ can be derived from $W_0(x)$, by dividing the interval (0, 1) in two equal halves, and writing '+' in the left half and '−' in the right half. $W_2(x)$ can be similarly obtained from $W_1(x)$ by dividing each of the two sub intervals of $W_1(x)$ into equal halves, and writing '+ −' in the sub intervals where it was '+' before, and '− +' where it was '−' before. $W_3(x)$ consists of four equal intervals, and can be derived from $W_0(x)$ by dividing the interval (0, 1) in two equal halves, and then writing '+ −' wherever the value was + in $W_0(x)$. Writing '+ −' over an interval implies that the interval is divided into two equal subintervals, and then '+' is put on the left subinterval, and '−' is put on the right subinterval. Similarly, writing '− +' over an interval implies that the interval is divided into two equal subintervals, and '−' is put on the left subinterval, and '+' is put on the right subinterval. In the same manner, $W_4(x)$ can be derived from $W_3(x)$, and $W_5(x)$ can be derived from $W_2(x)$. The choice of appropriate lower order from which a Walsh function of a higher order can be derived is decided by the number and location of zero crossings in the corresponding Walsh functions.

2.3   Classification of Walsh Functions of Various Orders in Self-similar Groups and Subgroups

The set of Walsh functions as a whole does not comprise of self-similar structures. However, these functions may be classified into different groups so that individual constituents of each group demonstrate self-similarity among themselves. Group I consists of Walsh functions of orders 1, 3, 7, 15, ... All of them are derived from $W_0(x)$. Group II consists of Walsh functions of orders 2, 4, 6, 8, 12, 14, ... It may be seen from Fig. 2.1 that each member of this group can be derived from members of group I. Similarly, Group III consists of Walsh functions of orders 5, 9, 11, 13, ... Figure 2.1 shows that they may be derived from Walsh functions of order 2, 6, 4 and 2 respectively. Each of the latter Walsh functions is a member of group II. Similarly, members of group IV are derived from members of group III, and so on. Figure 2.2 provides a genealogical chart of self-similar groups and subgroups in the set of unidimensional Walsh functions.

Group I is unique. It has no subgroups and members of the group portray self-similarity in their structures. All other groups have subgroups, and members of a particular subgroup exhibit self-similarity in their structures. Members of group II give rise to several subgroups. The order k of the first member of four of these subgroups, namely IIA, IIB, IIC and IID, belonging to group II, is within (0, 15). Each of these subgroups give rise to further subgroups. Taken together, all of them

Fig. 2.2 A genealogical chart of self-similar groups and subgroups in the set of Walsh functions

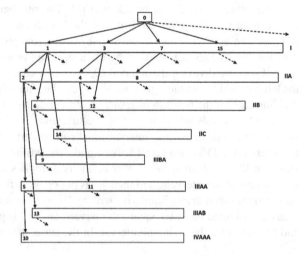

constitute group III, but individual subgroups are denoted by terms IIIAA, IIIAB, IIIAC, IIIBA, IIIBB, IIICA etc.

  In this nomenclature, the level of synthesis is indicated by the first roman numeral, followed by one or more letters of the alphabet to indicate the specific subgroup of the parent group from which it is generated; the last letter indicates the location of the first member of that specific subgroup with reference to the first member of other subgroups of the same level in an ascending order of sequence. If the orders ℓ_1, ℓ_2, ... of the first members of the subgroup are such that $\ell_1 < \ell_2 < \cdots$, then corresponding subgroups are denoted respectively by A, B, ... in the last letter of the notation. Members of a particular subgroup have distinct self-similarity in their structures. Table 2.1 shows the first few members of a few self-similar groups and subgroups. Each member is a Walsh function of a particular order. σth member of a specific group corresponds to a Walsh function of order $Z(\sigma)$ given by

$$Z(\sigma) = 2^{\sigma+b} + c, \quad \sigma = 1, 2, 3, \ldots \tag{2.1}$$

Table 2.1 Orders of the first few members of some of the self-similar groups and subgroups in the set of Walsh functions. For each of them the order of the Walsh function from which it is derived is given underneath in parenthesis

Group	Subgroup	Orders of the first few members of the subgroup					b	c	$Z(\sigma)$
I	–	1	3	7	15		0	−1	$2^{\sigma} - 1$
		(0)	(0)	(0)	(0)	...			
II	A	2	4	8			0	0	2^{σ}
		(1)	(3)	(7)	...	...			
	B	6	12				1	2^{σ}	$2^{\sigma+1} + 2^{\sigma}$
		(1)	(3)	...	...	...			
	C	14					2	$2^{\sigma+1} + 2^{\sigma}$	$2^{\sigma+2} + (2^{\sigma+1} + 2^{\sigma})$
		(1)	...	...	...	...			
	...	...	...	...	...	...	...	...	...
III	AA	5	11				1	$2^{\sigma} - 1$	$2^{\sigma+1} + (2^{\sigma} - 1)$
		(2)	(4)	...	...	...			
	AB	13					2	$2^{\sigma+1} + (2^{\sigma} - 1)$	$2^{\sigma+2} + [2^{\sigma+1} + (2^{\sigma} - 1)]$
		(2)	...	...	...	...			
	...	...	...	...	...	...	...	...	...
	BA	9					2	$2^{\sigma} - 1$	$2^{\sigma+2} + (2^{\sigma} - 1)$
		(6)	...	...	...	...			
	...	...	...	...	...	...	...	...	...
IV	AAA	10					2	2^{σ}	$2^{\sigma+2} + 2^{\sigma}$
		(5)	...	...	...	...			
	...	...	...	...	...	...	...	...	...

where b and c are two parameters specific for each subgroup. In general, b is a positive integer or zero, and c is either -1 or zero, or an integer whose value depends on the value of σ. For Group I and the first few subgroups of Group II and Group III, values of b, c, and $Z(\sigma)$ are given in the table. Our study is mostly limited to those subgroups, whose first member has order that falls within the range (0, 15), with occasional foray in Walsh functions of neighboring higher orders.

The list of groups and subgroups given in Table 2.1 is not exhaustive. The results presented here show only those groups and subgroups, for which the order of the first member falls within the range (0, 15). It is obvious that new subgroups will emerge when higher order Walsh functions are taken properly into account. For the groups and subgroups already identified, analytical expression (2.1) for $Z(\sigma)$ can be used to identify members lying among the higher orders of Walsh functions. More details on the self-similar groups and subgroups in the set of Walsh functions can be found in a recent publication [5].

2.4 Self-similarity in Radial Walsh Functions

The set of radial Walsh functions $\varphi_\ell(r)$ can be classified into distinct self-similar groups and subgroups as described in the previous section for the unidimensional Walsh functions. In $t(=r^2)$ space, the structure of radial Walsh functions $\varphi_\ell(t)$ becomes identical with unidimensional Walsh functions $W_\ell(x)$, as shown in Fig. 2.1.

2.5 Self-similarity in Annular Walsh Functions

The set of annular Walsh functions can also be classified into distinct self-similar groups and subgroups where members of each group possess self-similar structures or phase sequences. The classification scheme adopted for unidimensional or radial functions can be utilized directly for this case.

2.6 Radial and Annular Walsh Filters

Radial Walsh filters, derived from radial Walsh functions (as defined in Sect. 1.3.3 of Chap. 1), form a set of orthogonal phase filters that take on values either 0 or π phase, corresponding to +1 or -1 values of the radial Walsh functions over pre-specified annular regions of the circular filter [Note: $e^{i0} = +1$; $e^{i\pi} = -1$]. Order of these filters is given by the number of zero-crossings, or equivalently phase transitions within the domain over which the set is defined. In general, radial Walsh filters are binary phase filters or zone plates, each of them demonstrating distinct focusing characteristics. Annular Walsh filters are derived from the rotationally symmetric annular Walsh functions (as defined in Sect. 1.3.5 of Chap. 1) which form a complete set of orthogonal functions that take on values either +1 or −1 over the domain specified by the inner and outer radii of the annulus. The value of any annular Walsh function is taken as zero from the center of the circular aperture to the inner radius of the annulus. The three values 0, + 1 and −1 in an annular Walsh function can be realized in a corresponding annular Walsh filter by using transmission values of zero amplitude (i.e. an obscuration), unity amplitude and zero phase and unity amplitude and π phase respectively. Not only does the order of the annular Walsh filter, but also the size of the inner radius of the annulus provides an additional degree of freedom in tailoring of point spread function by using these filters for pupil plane filtering in imaging systems. These filters may be considered as ternary phase filters with values of transmission 0, + 1 and −1.

Since the transmission characteristics of kth order radial Walsh filter is uniquely specified by $\varphi_k(r)$, the kth order radial Walsh function, the kth order radial Walsh filter may be conveniently represented by the same notation. Similarly, the kth order annular Walsh filter with obscuration ratio ε is represented by the notation $\varphi_k^\varepsilon(r)$.

2.7   Self-similarity in Walsh Filters

2.7.1   Self-similarity in Radial Walsh Filters

The set of radial Walsh filters can be classified into distinct self-similar groups and subgroups where members of each subgroup possess self-similar structures or phase sequences, in the same manner as the radial Walsh functions from which they are derived. Group I is unique. It has no subgroup. Groups II, III and IV have subgroups as represented in Table 2.1. Members of each subgroup possess self-similar structures or phase sequences.

2.7.2   Self-similarity in Annular Walsh Filters

Annular Walsh filters may be classified into distinct self-similar groups and subgroups in the same manner as the annular Walsh functions from which they are derived. Group I is unique. It has no subgroups. Groups II, III and IV can be divided into subgroups as represented in Table 2.1. Members of each subgroup possess distinct self-similar structures or phase sequences.

References

1. Takayasu H (1990) Fractals in physical science. Manchester University, Manchester
2. Allain C, Cloitre M (1986) Optical diffraction on fractals. Phys Rev B 33:3566
3. Uozumi J, Asakura T (1994) Fractal optics. In: Dainty JC (ed) Current trends in optics. Academic Press, Cambridge, London, pp 189–196
4. Mandelbrot BB (1982) The fractal geometry of nature. Freeman, San Francisco
5. Mukherjee P, Hazra LN (2014) Self-Similarity in radial Walsh filters and axial intensity distribution in the Farfield diffraction pattern. J Opt Soc Am A 31(2):379–387
6. Harmuth HF (1972) Transmission of information by orthogonal functions. Springer, Berlin, pp 31

Chapter 3
Computation of Farfield Diffraction Characteristics of Radial and Annular Walsh Filters on the Pupil of Axisymmetric Imaging Systems

Abstract Pupil plane filtering by Walsh filters is a convenient technique for tailoring the intensity distribution of light near the focal plane of a rotationally symmetric imaging system. Walsh filters, derived from Walsh functions, form a set of orthogonal phase filters that take on values either 0 or π phase, corresponding to +1 or −1 values of the Walsh functions over prespecified annular regions of the circular filter. Order of these filters is given by the number of zero-crossings, or equivalently phase transitions within the domain over which the set is defined. In general, Walsh filters are binary phase zone plates, each of them demonstrating distinct focusing characteristics. With a backdrop of pupil plane filtering for image enhancement and a brief description of different types of zone plates for manipulation of axial intensity distribution, this chapter puts forward the inherent potential of Walsh filters in this context. The mathematical formulation utilized for computing the transverse and axial intensity distributions in and around the image/focal plane, when radial and annular Walsh filters are placed on the exit pupil of an axisymmetric imaging system, is presented.

Keywords Pupil plane filtering · Phase filters · Zone plates · Radial Walsh filters · Annular Walsh filters

3.1 Introduction

3.1.1 Pupil Plane Filtering and Walsh Filters

Pupil plane filtering for tailoring of the distribution of intensity on and around the focal/image plane of optical imaging systems is used extensively for image quality enhancement purposes. It has not only been used for increasing response of high frequency content of the image, but also for circumventing the Abbe/Rayleigh diffraction limit for resolution attainable by aberration free imaging systems. More than five decades ago Toraldo Di Francia [1] pioneered investigations on pupil plane filtering to enhance the resolving power of imaging system beyond the diffraction

© The Author(s) 2018

L. Hazra and P. Mukherjee, *Self-similarity in Walsh Functions and in the Farfield Diffraction Patterns of Radial Walsh Filters*, SpringerBriefs in Applied Sciences and Technology, DOI 10.1007/978-981-10-2809-0_3

limit. He observed that a set of amplitude and/or phase filters on the pupil of an image forming system can achieve the same albeit over a restricted field [2, 3]. He described superresolving filters that use array of annular elements. Lohmann [4], Kartashev [5], Lukosz [6], Frieden [7, 8], Boyer and coworkers [9–11] and Boivin and Boivin [12–14] carried out further investigation on pupil plane filters with an aim to exceed the classical resolution limit. Boivin and Boivin presented five different filter designs with chosen positions for zeroes in intensity in the focal plane. Cox et al. [15] studied five different filters consisting of arrays of contiguous annuli of equal area and varying real values of amplitude transmission with the zeroes chosen to be the same as Boivin and Boivin. Hegedus and Sarafis [16] highlighted the potential of these filters in scanning microscopy. In most of these studies, the filter consists of variations of amplitude transmittance or variation in complex amplitude among the zones. In case of the latter, the studies were mostly limited to a π-phase shift between adjacent zones. With the advent of diffractive optics technology, Sales and Morris [17] explored the use of multiphase structures. In these structures the phase transmission of each zone can assume an arbitrary value in the interval $(0, 2\pi)$. Castaneda et al. [18] proposed annular apodizers or shade masks that increase the depth of focus and reduce the influence of spherical aberrations. In the recent past extensive studies have been carried out on the analysis and synthesis of lossy (only amplitude), lossless (only phase) and leaky (hybrid) filters to improve not only transverse superresolution but for improvement of axial sectioning and axial superresolution [19–44]. Shepard et al. [45] investigated the focal distribution for Toraldo filters improving three dimensional superresolution. A novel procedure to design both axial and transverse superresolving filters were proposed by de Juana et al. [46], and three dimensional superresolution by three zone complex pupil filters [47] and by annular binary filters [48] has also been proposed.

Pupil plane filters allow complete reshaping of the spatial intensity distribution near the focus [49, 50] to cater to apodization or 3D superresolution. Annular pupil-plane masks [51] may be used to control light intensity near the focus and finds application in microscopy and optical tweezers. Synthesis of three dimensional light fields have also been done using micro structures on diffractive optical elements [52] which could find application in lithography where depth of focus is as important as resolution and in volume memories or near field optics. Pupil engineering to create sheets, lines, and multiple spots at the focal region is also being investigated vigorously [53]. Different types of amplitude and/or phase filters have already been investigated often with repetition of efforts in different regions of the electromagnetic spectrum. Except for trivial cases, no closed form analytical solution is available for solving the inverse problem of synthesis of the optimum pupil plane filter catering to a prespecified diffraction pattern, and in practice one needs to take recourse to semi numerical algorithms for tackling the problem at hand. Usually these techniques involve a decomposition of the pupil function in a finite set of base functions whose diffraction patterns are known or can be computed conveniently. Problems regarding appropriate size of the set, convergence etc. can be somewhat reduced by using a set of orthogonal functions as the base functions. The search for a suitable set of base functions has led to the use of Straubel functions, Bessel functions, Zernike circle polynomials, prolate spheroidal wave functions and Tchebycheff polynomials among others [9, 54–60]. Most of these functions have provided useful solutions as sought, but none of them yet emerged as a panacea for the diffraction problems in general.

With this backdrop in view the convenient amenability of Walsh functions, a complete set of orthogonal functions as base functions have been proposed [61]. Walsh filters derived from Walsh functions or more precisely annular Walsh filters were used as pupil plane filters for tailoring the resolution of microscopic imaging by Hazra [62].

Radial Walsh filters [63] were developed for the azimuth invariant case for tackling problems of apodization and adaptive optics [61, 64]. Diffraction properties of Walsh filters on the transverse farfield plane were also studied in the context of tailoring of resolution in microscopic imaging [62]. The farfield amplitude characteristics of some of these filters have been studied [66] to underscore their potential for effective use in several demanding applications like high resolution microscopy, optical data storage, microlithography, optical encryption and optical micromanipulation [67–73]. Not only for obvious energy considerations, but also for their higher inherent potential in delivering complex farfield amplitude distributions, annular phase filters are being investigated. A systematic study on the use of phase filters on annular pupils can be conveniently carried out with the help of annular Walsh filters derived from the annular Walsh functions.

Ready availability of high efficiency spatial light modulators has facilitated the practical realization of these energy efficient phase filters [74]. Nevertheless, it is obvious that, in general, practical implementation of filters with binary or ternary phase values is relatively easier than filters with continuously varying phase. Use of Walsh function based analysis and synthesis of pupil plane filters has become useful, for it can circumvent the tricky problem of synthesizing continuously varying phase by providing alternative analytical treatments that can obviate the need for any ladder step approximation of a continuous variation, and provide an approach that can directly deal with finite number of discrete phase levels.

3.1.2 Zone Plates and Walsh Filters

There has been a resurgence of interest in zone plates [75] during the last years since they are being used as key imaging elements in several scientific and technological areas. The Fresnel zone plate acts as a lens that forms an image by diffraction rather than by refraction as in a normal lens. The basic properties of these lenses have been known for more than a century [76–78]. Since the diffraction efficiency [79, 80] achievable with conventional zone plates is far below the minimum acceptable level for applications in the visible and infrared regions of the electromagnetic spectrum, the practical applications of zone plates in these regions of the spectrum have been limited. But in the ultraviolet and soft X-ray regions of the spectrum, zone plates find many practical applications [81–85] due to unavailability of suitable refracting materials. Of particular mention among their numerous applications are those in microanalysis and X-ray microscopy, plasma diagnostics, synchrotron radiation focusing [82–90], imaging and focusing of atoms [91], terahertz tomography, high energy astronomy, ophthalmology and soft X-ray microscopy [92–96].

With its alternating transparent and opaque zones whose radii are proportional to square root of natural numbers, these zone plates produce simultaneously multiple images, both real and virtual [97, 98]. One of the main shortcomings of Fresnel zone plates is their high chromatic aberration. A new class of zone plates called fractal zone plates forms a recent area of research [99–104] as they can improve the performance of classical Fresnel zone plates in certain applications where multiple foci are needed. Fractal zone plates provide an extended depth of field and a reduction in chromatic aberration when used in incoherent imaging using poly-chromatic light in the visible region [105].

A fractal zone plate can be thought as a conventional zone plate with certain missing zones. These zone plates have fractal structures. The axial irradiance exhibited by these fractal zone plates have self-similarity properties that can be correlated to that of the diffracting aperture. The axial irradiance provided by a fractal zone plate when illuminated with parallel wave fronts present multiple foci, the main lobe of which coincide with those of the associated conventional zone plates. However, the internal structure of each focus exhibits a characteristic self-similar fractal structure, reproducing the self-similarity of the original fractal zone plate [103]. If the pupil function holds a fractal structure, it is well known that the Fourier transform preserves fractal properties and then may be inferred that such pupil will produce irradiance along the optical axis also with a fractal profile [102, 106]. Studies on the focusing properties of fractal zone plates reveal that the axial irradiance along the optical axis produced by these pupils exhibited the self-similarity of the fractal zone plates or showed a characteristic fractal profile. These fractal zone plates were mainly generated from a triadic Cantor set [99]. Saavedra et al. [100] studied pupils which possessed fractal structure along the square of the radial co-ordinate and were considered as conventional zone plates with some missing clear zones. Fractal zone plates derived from Cantor fractals with variable lacunarity and its effects on the Fraunhofer diffraction pattern were also studied [101, 102]. The interesting feature of axial self-similarity produced by these fractal zone plates were called the axial scale property.

Fractal square zone plates based on the triadic Cantor set was introduced by Calatayud et al. [103] They used the concept of coding the Cantor set using an array of binary elements. Fractal square zone plates are zone plates with a fractal dis-tribution of square zones. The axial and transverse irradiance produced by the fractality of these diffractive lenses were numerically evaluated. Cantor dust zone plates based on the two dimensional (2D) fractal structure was introduced by Monsoriu et al. [104]. It was geometrically constructed from a 2D Cantor set. Mathematically, the Cantor dust was represented by a square array of binary ele-ments. Cantor dust zone plates can be constructed as a photon sieve constituted by an array of rectangular transparent holes with squares distributed along the square transverse co-ordinates. They preserve the characteristic self-similar behavior of a fractal square zone plate in their axial irradiance pattern. Under monochromatic illumination, a Cantor dust zone plate gives rise to a focal volume containing a delimited sequence of two-arm-cross patterns which are axially distributed according to the self-similarity of the lens. One potential application of Cantor dust zone plate is to generate a reference pattern in optical alignment and calibration of 3D systems. Fractal zone plates are diffractive lenses characterized by the fractal structure of their foci. The one dimensional binary function associated with a member of the Cantor set developed up to third stage was used. A change of variables and rotation around the origin resulted in a circularly symmetric fractal zone plate and white light imaging with such zone plates were studied [105]. Fractal generalized zone plates which form a set of periodic diffractive optical elements with circular symmetry were introduced [107]. This allows increase in the number of foci of a conventional fractal zone plate keeping self-similarity within axial irradiance. Fractal photon sieves [108] are photon sieves in which the pin holes are appropriately distributed over the zones of a fractal zone plate. Compared with a conventional photon sieve, fractal photon sieves exhibit an extended depth of field and reduced chromatic aberration. The fractal photon sieves potentially improve the performance of fractal zone plates.

Fractal zone plates find application in scientific and technological areas where conventional zone plates have been successfully applied. Modified fractal generalized zone plates extend depth of focus and has been used in spectral domain optical coherence tomography [109]. Fractal zone plates also inspired the invention of other photonic structures [109–111]. Particularly recent proposals of optical tweezers use phase filters to facilitate the passing of particles in three dimensional structure [104]. Fractal generalized zone plates and spiral fractal zone plates [109] finds application in trapping and optical micromanipulation [112–118]. Non uniform distribution of fractal zone plate's focal points along the optical axis could be exploited in the design of multi focal contact lenses for correction of presbyopia. Fractal zone plates can also be used in other regions of the electromagnetic spectrum such as microwaves and X-rays and even with slow neutrons.

In recent years there has been considerable interest in the study of quasi-periodic and fractal optical elements. Fibonacci fiber Bragg gratings can transform evanescent waves into propagating waves for farfield superresolution imaging. Fibonacci sequences are being experimented in multilayer and linear gratings, circular gratings and spiral zone plates. Fibonacci lenses are being used to generate arrays of optical vortices to trap micro particle for driving optical pumps [119–128]. Fibonacci diffraction gratings [129, 130] which is an archetypal example of aperiodicity and self-similarity are finding application as image forming devices. Current state-of-the-art on different aspects of these aperiodic zone plates can be found in references [131–133].

Radial and Annular Walsh filters of different orders possess an inherent self-similar structure [134, 135]. From lower to higher orders, these orders which have the self-similar replicating phase sequences can be clubbed into a group. As a whole, the set of radial and annular Walsh functions does not comprise of self-similar structures. But, as enunciated in the earlier chapter, distinct groups of radial and annular Walsh functions demonstrate self-similarity among their individual constituents. Groups are divided into subgroups and members of each subgroup can be defined analytically. The axial and transverse intensity distributions [134–136] exhibited by each group member have self-similar properties as elaborated in the subsequent chapters. They correspond to the structural self-similarity of the filters in the group, a characteristic quite similar to that exhibited by fractal zone plates and other fractal diffracting apertures. Radial and Annular Walsh filters may be used as pupil plane filters for tailoring the axial, transverse and three dimensional intensity distribution near the focal plane of a rotationally symmetric imaging system. Annular Walsh filters may also be considered as annular versions of generalized zone plates or ternary phase plates with self-similarity. It is obvious that the obscuration ratio of these annular filters provide an additional and highly useful degree of freedom in tailoring of farfield diffraction patterns.

The next section presents the mathematical formulation for the transverse and axial intensity distribution on the farfield when radial and annular Walsh filters are used as pupil plane filters in axisymmetric imaging systems. Our formulation enables computation of normalized transverse/axial intensity distributions, on and around the focal/image plane, with good accuracy by taking recourse to special numerical quadrature techniques, described below.

The transverse amplitude distribution in the farfield diffraction pattern of the exit pupil of an optical imaging system is obtained on the transverse focal/image plane of the system, and is expressed mathematically as the two dimensional Fourier transform of the pupil function. In case of axisymmetric imaging systems, the latter reduces to the Hankel transform of the azimuth invariant pupil function. The transverse intensity distribution is given by the squared modulus of the corresponding amplitude distribution. Determination of transverse intensity distributions on other planes neighboring the image/focal plane call for use of the theory of Fresnel diffraction. Mathematically, the evaluation of transverse intensity distributions on other planes neighboring the image/focal plane reduces to determination of squared modulus of the two dimensional Fourier transform of a defocused pupil function. Again, for axisymmetric imaging systems, computational problem boils down to evaluation of the Hankel transform of a defocused azimuth invariant pupil function. The latter has been tackled by using special numerical quadrature techniques mentioned above.

By means of suitable scaling and normalization, intensity distributions are presented in terms of dimensionless variables a and p for the axial distance and the transverse distance respectively. The numerical results, thereby, retain their practical usefulness for optical imaging systems with different values for numerical aperture and operating wavelengths. Our analytical treatment is based on scalar diffraction theory, as is customary for a wide class of imaging systems used for optical instrumentation. Therefore, this analysis retains its validity for systems with small to moderate values of numerical aperture, for in the case of systems with very high values of numerical aperture, analytical treatments need to be based on vector diffraction theory to take proper account of polarization effects. Also, our treatment retains its validity so long as the 'critical dimension' of the diffracting structure, e.g.

zone width of the outermost zone of radial or annular Walsh filters, is greater than, say 5λ, where λ is the operating wavelength.

3.2 Farfield Diffraction Characteristics of Radial and Annular Walsh Filters: Mathematical Formulation

The complex amplitude distribution on the transverse image plane located at O' in the farfield of an axially symmetric imaging system (Fig. 3.1) is given by the two dimensional Fourier transform of the complex amplitude distribution over the exit pupil. For rotationally symmetric imaging systems, amplitude distribution on the image plane becomes azimuth invariant, and can be expressed to a multiplicative constant m as Hankel transform of the pupil function, and it may be expressed as [137–139]

$$F(p) = m \int_0^1 f(r) J_0(pr) r \, dr \qquad (3.1)$$

where $f(r)$ is the azimuth invariant pupil function representing the complex amplitude distribution over the exit pupil at E'. r is the normalized radial distance of a point A on the exit pupil from the optical axis. Let $AE' = \rho$. $r = \rho/\rho_{max}$ where ρ_{max} is the radius of the exit pupil. p is the reduced diffraction variable defined as

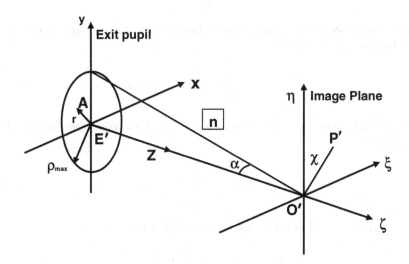

Fig. 3.1 Exit pupil and farfield plane in the image space of an axially symmetric imaging system

$$p = \frac{2\pi}{\lambda}(n \sin \alpha)\chi \qquad (3.2)$$

where $\chi(= O'P')$ is the geometrical distance of a point P' on the image from the centre of the diffraction pattern. $(n \sin \alpha)$ is the the image space numerical aperture. k, the propagation constant, is equal to $(\frac{2\pi}{\lambda})$, where λ is the operating wavelength. n is the image space refractive index.

3.3 Radial Walsh Filters on the Exit Pupil

3.3.1 Transverse Intensity Distribution on the Focal/Image Plane

For an aberration free pupil without obscuration having uniform amplitude, pupil function $f(r)$ is expressed as

$$
\begin{aligned}
f(r) &= 1 \text{ over the aperture} \\
&= 0 \text{ elsewhere}
\end{aligned} \qquad (3.3)
$$

For this pupil, the amplitude at the center of the diffraction pattern is

$$F(0) = m \int_0^1 r\,dr = \frac{m}{2} \qquad (3.4)$$

The normalized complex amplitude distribution on the transverse plane with pupil function $f(r)$ is given by

$$F_N(p) = \frac{F(p)}{F(0)} = 2 \int_0^1 f(r)J_0(pr)r\,dr \qquad (3.5)$$

For the kth order radial Walsh filter on the pupil, the pupil function may be expressed as

$$f(r) = \varphi_k(r) = \sum_{m=1}^{M} e^{ik\psi_m} B_m(r) \qquad (3.6)$$

where $B_m(r)$ are radial Walsh Block functions defined as

$$
\begin{aligned}
B_m(r) &= 1; \quad r_{m-1} \le r \le r_m \\
&= 0; \quad \text{otherwise}
\end{aligned} \qquad (3.7)
$$

where

$$r_m = \left[\frac{m}{M}\right]^{\frac{1}{2}} \quad \text{and} \quad r_{m-1} = \left[\frac{m-1}{M}\right]^{\frac{1}{2}} \tag{3.8}$$

and $M = 2^\gamma$ is the largest power of 2 that just exceeds k. Over the mth zone of $\varphi_k(r)$, value of $k\psi_m$ is 0 or π, if $\varphi_k(r) = +1$ or -1 respectively.

For better accuracy of results each of the M zones of the pupil is divided into J number of equal area subzones such that the whole exit pupil has $T = MJ$ number of equal area concentric zones (Fig. 3.2). The outer and inner radii of the jth subzone of the mth zone are given by

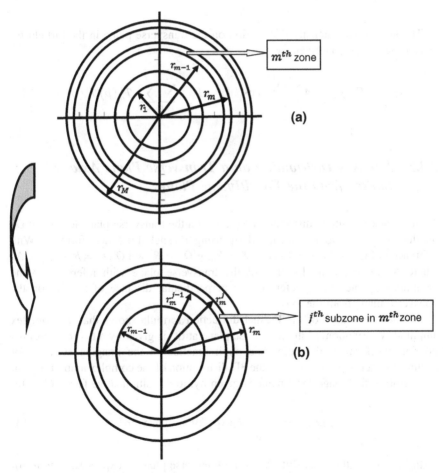

Fig. 3.2 **a** mth zone in a phase filter consisting of M concentric equal area zones. **b** jth subzone in the mth zone in a phase filter consisting of J equal area subzones in the mth zone

$$r_m^j = \left[\frac{(m-1)J + j}{T}\right]^{\frac{1}{2}} \quad \text{and} \quad r_m^{j-1} = \left[\frac{(m-1)J + (j-1)}{T}\right]^{\frac{1}{2}} \text{respectively.} \quad (3.9)$$

The normalized complex amplitude distributions on transverse plane in the farfield for Walsh filters $\varphi_\ell(r)$ is given by

$$F_N(p) = 2\sum_{m=1}^{M} \exp[ik\psi_m] \sum_{j=1}^{J} \int_{r_m^{j-1}}^{r_m^j} J_0(pr)rdr = 2\sum_{m=1}^{M} \exp[ik\psi_m] \sum_{j=1}^{J} \mathcal{J}_m^j(p) \quad (3.10)$$

where

$$\mathcal{J}_m^j(p) = \left[\frac{r_m^j J_1(pr_m^j) - r_m^{j-1} J_1(pr_m^{j-1})}{p}\right] \quad (3.11)$$

The normalized intensity distribution on the transverse plane in the farfield for Walsh filters $\varphi_\ell(r)$ is given by

$$I_N(p) = |F_N(p)|^2 = 4\sum_{m=1}^{M}\sum_{u=1}^{M} \cos[k\psi_m - k\psi_u] \sum_{j=1}^{J}\sum_{v=1}^{J} \mathcal{J}_m^j(p)\mathcal{J}_u^v(p) \quad (3.12)$$

3.3.2　Intensity Distribution on a Transverse Plane Axially Shifted from the Focal/Image Plane

The complex amplitude distribution $F(\chi, \Delta\zeta)$ on the transverse plane located at O'' on the axis needs to be calculated by using Fresnel diffraction formula. With reference to Figs. 3.3 and 3.4, let $E'O' = R_o'$, $E'O'' = R'$ and $O'O'' = R' - R_o' = \Delta\zeta$ where $\Delta\zeta$ refers to the location of the transverse plane with reference to the focal/image plane, and χ refers to the perpendicular distance of a point on the transverse plane from the axis.

Detailed derivation of the Fresnel diffraction formula shows that the complex amplitude distribution on the transverse plane at O'' is given by a two dimensional Fourier transform of the defocused pupil function of the imaging system. For axially symmetric pupil function, the 3D distribution of the complex amplitude near the region of the image/focal plane is given by, to a multiplicative factor C, [140]

$$F(\chi, \Delta\zeta) = C \int_0^1 f(r)\exp[ikW_{20}r^2] J_0(pr)rdr \quad (3.13)$$

In Eq. (3.13), the axial shift $\Delta\zeta$ of the transverse plane is expressed in terms of 'defocus' aberration W_{20}, where the wave aberration function $W(r)$ is given by

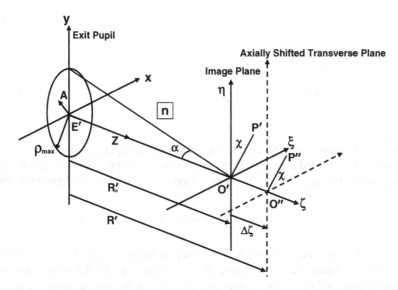

Fig. 3.3 Exit pupil, image plane and an axially shifted transverse plane in the image space of an axially symmetric imaging system

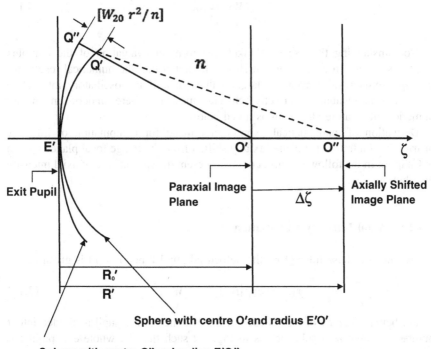

Fig. 3.4 Shift of image plane by $\Delta\zeta$ may be represented equivalently by defocus or 'Depth of Focus' aberration $[Q''Q']$

$$W(r) = W_{20}r^2 \qquad (3.14)$$

W_{20} is related to $\Delta\zeta$ by [141].

$$W_{20} = \frac{1}{2} n \rho'^2_{max} \left(\frac{1}{R'_o} - \frac{1}{R'} \right) = \frac{1}{2n} (n \sin \alpha)^2 \Delta\zeta \qquad (3.15)$$

Instead of actual distances χ and $\Delta\zeta$, expressions for complex amplitude $F(p, a)$ are derived for normalized dimensionless parameters p and a respectively. p has been already defined in Eq. (3.2) and the reduced axial co-ordinate a is related to $\Delta\zeta$ by

$$a = \frac{1}{n\lambda} (n \sin \alpha)^2 \Delta\zeta \qquad (3.16)$$

where $\Delta\zeta$ is the actual longitudinal shift representing the axial distance of the transverse plane from the paraxial focal plane as shown in Fig. 3.3, n represents the refractive index of the image space, and $(n \sin \alpha)$ is the image space numerical aperture.

W_{20} is related to a by

$$kW_{20} = a\pi \qquad (3.17)$$

Notwithstanding the use of Hankel transform in evaluation of the complex amplitude on a defocused plane, it should be noted that the numerical results for intensity remain valid even for larger values of a if the oscillatory integral in Eq. (3.13) is evaluated correctly. Special measures were undertaken in our numerical quadrature algorithm as noted below.

Derivation of mathematical expression utilized for computation of intensity distribution on transverse planes axially shifted from the image/focal plane is given in Chap. 6. In the following subsection we present the special case of axial intensity distribution.

3.3.2.1 Axial Intensity Distribution

In presence of phase filter $Q(r)$, the defocused pupil function $f(r)$ is given by

$$f(r) = \exp\left[ik\{Q(r) + W_{20}r^2\} \right] \qquad (3.18)$$

For better accuracy of results each of the M zones of the pupil is divided into J number of equal area subzones as in Fig. 3.2 such that the whole exit pupil has $T = MJ$ number of equal area concentric zones. For each of these sub zones the outer and inner radii r_m^j and r_m^{j-1} are given by

$$r_m^j = \left[\frac{(m-1)J+j}{T}\right]^{\frac{1}{2}} \quad \text{and} \quad r_m^{j-1} = \left[\frac{(m-1)J+(j-1)}{T}\right]^{\frac{1}{2}} \tag{3.19}$$

The normalized complex amplitude at a point O'' on the axis is [65]

$$F_N(a) = F_N(\Delta\zeta) = 2\sum_{m=1}^{M} \exp[ik\psi_m] \sum_{j=1}^{J} \exp[ik\bar{W}_m^j] \int_{r_m^{j-1}}^{r_m^j} rdr \tag{3.20}$$

The axial shift $O'O'' = \Delta\zeta$ (as in Fig. 3.3). $\bar{W}_m^j$ is the average value of $W(r)$ over the jth subzone of the mth zone and is given by

$$\bar{W}_m^j = \frac{\int_{r_m^{j-1}}^{r_m^j} W(r)rdr}{\int_{r_m^{j-1}}^{r_m^j} rdr} = \frac{W_{20}}{2T}[2(m-1)J+(2j-1)] \tag{3.21}$$

and

$$\int_{r_m^{j-1}}^{r_m^j} rdr = \Re_j = \left[\frac{\left(r_m^j\right)^2}{2} - \frac{\left(r_m^{j-1}\right)^2}{2}\right] = \left[\frac{1}{2T}\right] \tag{3.22}$$

for $k\psi_m = 0$ or π, values of $\exp(ik\psi_m)$ are $+1$ or -1 respectively.
From (3.20)–(3.22)

$$F_N(a) = \frac{1}{T}\sum_{m=1}^{M} \exp[ik\psi_m] \sum_{j=1}^{J} \exp[ik\bar{W}_m^j] \tag{3.23}$$

The normalized intensity distribution along the axis may be expressed in terms of a as

$$I_N(a) = |F_N(\Delta\zeta)|^2 = \frac{1}{T^2}\sum_{m=1}^{M}\sum_{u=1}^{M}\sum_{j=1}^{J}\sum_{v=1}^{J} \cos\left[\{k\psi_m - k\psi_u\} + \{k\bar{W}_m^j - k\bar{W}_u^v\}\right] \tag{3.24}$$

3.4 Annular Walsh Filters

3.4.1 Transverse Intensity Distribution

It has been noted in the earlier section that the complex amplitude distribution on the image plane corresponding to an axial object point is given by the farfield diffraction pattern of the pupil function over the exit pupil. The complex amplitude $F(p)$ at a point P′ in the farfield due to a point object on the axis is, apart from the multiplicative constant, given by the Hankel transform of order zero of the pupil function

$$F(p) = \int_0^1 f(r)J_0(pr)r dr \tag{3.25}$$

For a specific central obscuration ratio ε, annular Walsh filters of various orders can be obtained from the corresponding annular Walsh functions by realizing transmission values $+1$ and -1 by zero and π phase filters respectively. In presence of an annular Walsh filter on a uniform pupil, the pupil function $f(r)$ is given by

$$f(r) = \begin{cases} 0, & 0 \le r < \varepsilon \\ \varphi_k^\varepsilon(r), & \varepsilon \le r < 1 \end{cases} \tag{3.26}$$

It may be noted that $f(r)$ is binary (value either 0 or $+1$) only in the case of zero order annular Walsh function $\varphi_0^\varepsilon(r)$, for all other orders $f(r)$ is ternary with value either 0, $+1$ or -1.

With a phase filter $Q(r)$ on the exit pupil of the imaging system the pupil function $f(r)$ may be expressed as

$$f(r) = \exp[ikQ(r)] \tag{3.27}$$

For a phase filter consisting of M concentric equal area annular zones over the annular aperture,

$$\begin{aligned} kQ(r) &= \sum_{m=1}^M k\psi_m B_m(r), & \varepsilon \le r \le 1 \\ &= 0, & 0 \le r < \varepsilon \end{aligned} \tag{3.28}$$

Phase over the mth zone is $k\psi_m$. $B_m(r)$ are annular Walsh Block functions defined as

$$\begin{aligned} B_m(r) &= 1; & r_{m-1} \le r \le r_m \\ &= 0; & \text{otherwise} \end{aligned} \tag{3.29}$$

where

$$r_m = \sqrt{\frac{[(M-m)\varepsilon^2 + m]}{M}} \tag{3.30}$$

ε is the central obscuration ratio.

Note that over the annular aperture, $\varepsilon \le r \le 1$, $\varphi_k^\varepsilon(r)$ takes values either $+1$ or -1, that correspond to values 0 or π for $kQ(r)$ respectively.

The normalized farfield amplitude pattern $F_N^\varepsilon(p)$ of an annular Walsh filter $\varphi_h^\varepsilon(r)$ is given by

$$F_N^\varepsilon(p) = 2\int_\varepsilon^1 \varphi_k^\varepsilon(r)J_0(pr)rdr = 2\sum_{m=1}^M h_{km}\mathcal{J}_m(p) \qquad (3.31)$$

where

$$\mathcal{J}_m(p) = \int_\varepsilon^1 B_m(r)J_0(pr)rdr = \left[\frac{r_m^2 J_1(pr_m)}{pr_m} - \frac{r_{m-1}^2 J_1(pr_{m-1})}{pr_{m-1}}\right] \qquad (3.32)$$

h_{km} are the elements of a $(M \times M)$ Hadamard matrix whose elements are $+1$ or -1. The elements of the Hadamard matrix are determined by the governing relation (1.16) for annular Walsh functions.

The normalized transverse intensity on the farfield plane is given by

$$I_N^\varepsilon(p) = \left|F_N^\varepsilon(p)\right|^2 \qquad (3.33)$$

3.4.2 Axial Intensity Distribution

The normalized intensity at a point O'' on an axially shifted image plane (Fig. 3.3) with an annular Walsh filter on the exit pupil is given by

$$I_N^\varepsilon(a) = \frac{1}{T^2}\sum_{m=1}^M\sum_{u=1}^M\sum_{j=1}^J\sum_{v=1}^J \cos\left[\{k\psi_m - k\psi_u\} + \{k\bar{W}_m^j - k\bar{W}_u^v\}\right] \qquad (3.34)$$

where the axial shift $O'D' = \Delta\zeta$ is the actual longitudinal shift representing the axial distance of the axially shifted transverse plane from the paraxial focal plane. $\bar{W}_m^j$ is the average value of $W(r)$ over the jth subzone as given by Eq. (3.21). M represents the number of zones into which the pupil is divided. Phase over the mth zone is $k\psi_m$. For $k\psi_m = 0$ or π, values of $\exp(ik\psi_m)$ are $+1$ or -1 respectively.

For better accuracy of results each of the M zones of the pupil is divided into J number of equal area subzones as in Fig. 3.5 such that the whole exit pupil has $T = MJ$ number of equal area concentric zones. For each of these sub zones the outer and inner radii r_m^j and r_m^{j-1} are given by

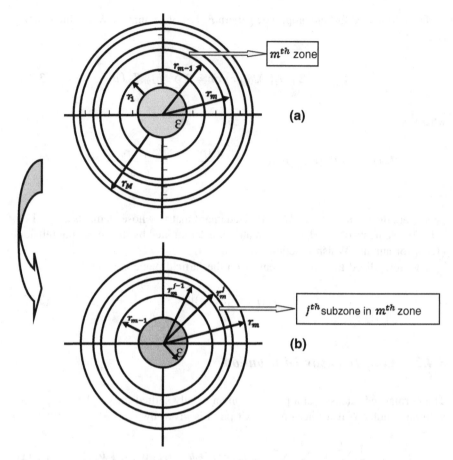

Fig. 3.5 **a** mth zone in a phase filter consisting of M concentric equal area zones. **b** jth subzone in the mth zone in a phase filter consisting of J equal area subzones in the mth zone. ε is the central obscuration ratio of the annular Walsh filter

$$r_m^j = \varepsilon + \left[\frac{(m-1)J+j}{T}\right]^{\frac{1}{2}} \quad (3.35)$$

and

$$r_m^{j-1} = \varepsilon + \left[\frac{(m-1)J+(j-1)}{T}\right]^{\frac{1}{2}} \quad (3.36)$$

where ε represents the central obscuration ratio of the annular Walsh filter.

References

1. Toraldo di Francia G (1958) La Diffrazione Delle Luce, Edizioni Scientifiche Einaudi
2. Toraldo di Francia G (1952) Super-gain antennas and optical resolving power. Nuovo Cimento Supple 9(3):426–438
3. Toraldo di Francia G (1952) Nuovo pupille superresolventi. Atti Fond Giorgio Ronchi 7:366–372
4. Brown BR, Lohmann AW (1966) Complex spatial filtering with binary marks. Appl Opt 5 (6):967–969
5. Kartashev AI (1960) Optical systems with enhanced resolving power. Opt Spectrosc 9:204–206
6. Lukosz W (1966) Optical systems with resolving power exceeding the classical limit. J Opt Soc Am 56(11):1463–1472
7. Frieden BR (1969) On arbitrarily perfect imaging with a finite aperture. Opt Acta 16 (6):795–807
8. Frieden BR (1971) VIII Evaluation, design and extrapolation methods for optical signals based on use of the prolate functions. In: Wolf E (ed), North Holland. Prog Opt 9:311–407
9. Boyer GR, Sechaud M (1973) Superresolution by Taylor filters. Appl Opt 17(4):893–894
10. Boyer GR (1976) Pupil filters for moderate superresolution. Appl Opt 15(12):3089–3093
11. Boyer GR (1983) Realisation d'un filtrage super-resolvant. Opt Acta 30:807–816
12. Boivin R, Boivin A (1980) Optimized amplitude filtering for superresolution over a restricted field. I. Achievement of maximum central irradiance under an energy constraint. Opt Acta 27(5):587–610
13. Boivin R, Boivin A (1980) Optimized amplitude filtering for superresolution over a restricted field. II. Application of the impulse generating filter. Opt Acta 27:1641–1670
14. Boivin R, Boivin A (1983) Optimized amplitude filtering for superresolution over a restricted field. III. Effects due to variation of the field extent. Opt Acta 30:681–688
15. Cox IJ, Sheppard CJR, Wilson T (1982) Reappraisal of arrays of concentric annuli as superresolving filters. J Opt Soz Am 72(9):1287–1291
16. Hegedus ZS, Sarafis V (1986) Superresolving filters in confocally scanned imaging systems. J Opt Soc Am A 3(11):1892–1896
17. Sales TRM, Morris GM (1997) Diffractive superresolution elements. J Opt Soc Am A 14 (7):1637–1646
18. Ojeda-Castaneda J, Andrés P, Diaz A (1986) Annular apodizers for low sensitivity to defocus and to spherical aberration. Opt Lett 11(8):487–489
19. Sheppard CJR, Hegedus ZS (1988) Axial behavior of pupil plane filters. J Opt Soc Am A 5 (5):643–647
20. Hazra LN (1977) A new class of optimum amplitude filters. Opt Commun 21(2):232–236
21. Ding H, Li Q, Zou W (2004) Design and comparison of amplitude-type and phase-only transverse super-resolving pupil filters. Opt Commun 229(1–6):117–122
22. Sheppard CJR, Sharma MD, Arbouet A (2000) Axial apodizing filters for confocal imaging. Optik 111(8):347–354
23. Yun M, Wang M, Liu L (2006) Transverse superresolution with the radial continuous transmittance filter. Optik 117(5):240–245
24. Sales TRM, Morris GM (1998) Axial superresolution with phase-only pupil filters. Opt Commun 156(4–6):227–230
25. Martinez-Corral M, Caballero MT, Stelzer EHK, Swoger J (2002) Tailoring the axial shape of the point spread function using the Toraldo concept. Opt Express 10(1):98–103
26. Luo H, Zhou C (2004) Comparison of superresolution effects with annular phase and amplitude filters. Appl Opt 43(34):6242–6247
27. Liu X, Liu L, Liu D, Bai L (2006) Design and application of three-zone annular filters. Optik 117(10):453–461

28. Sheppard CJR, Campos J, Escalera JC, Ledesma S (2008) Two-zone pupil filters. Opt Commun 281:913–922
29. Sheppard CJR, Campos J, Escalera JC, Ledesma S (2008) Three-zone pupil filters. Opt Commun 281:3623–3630
30. Martinez-Corral M, Andrés P, Ojeda-Castaneda J (1994) On-axis diffractional behavior of two dimensional pupils. Appl Opt 33(11):2223–2229
31. Martinez-Corral M, Andrés P, Ojeda-Castaneda J, Saavedra G (1995) Tunable axial superresolution by annular binary filters. Application to confocal microscopy. Opt Commun 119(5–6):491–498
32. Ledesma S, Campos J, Escalera JC, Yzuel MJ (2004) Simple expressions for performance parameters of complex filters, with applications to super-Gaussian phase filters. Opt Lett 29 (9):932–934
33. Ledesma S, Escalera JC, Campos J, Yzuel MJ (2005) Evolution of the transverse response of an optical system with complex filters. Opt Commun 249(1–3):183–192
34. Jabbour TG, Petrovich M, Kuebler SM (2008) Design of axially super resolving phase filters using the method of generalized projection. Opt Commun 281(8):2002–2011
35. Sheppard CJR (2007) Fundamentals of superresolution. Micron 38:165–169
36. Martínez-Corral M, Saavedra G (2009) The resolution challenge in 3D optical microscopy. In: Wolf E (ed). Prog Opt 53:1–67
37. Hazra LN, Reza N (2010) Optimal design of Toraldo super resolving filters. In: Procedings of SPIE 7787, Novel Optical Systems Design and Optimization XIII, 77870D
38. Sheppard CJR (2011) Binary phase filters with a maximally flat response. Opt Lett 36 (8):1386–1388
39. Leizerson I, Lipson SG, Sarafis V (2002) Superresolution in far-field imaging. J Opt Soc Am A 19(3):436–443
40. Martinez-Corral M, Ibáñez-López C, Caballero MT, Saavedra G (2003) Axial gain resolution in optical sectioning fluorescence microscopy by shaded-ring filters. Opt Express 11(15):1740–1745
41. Hegedus ZS (1985) Annular pupil arrays—application to confocal scanning. Opt Acta 32 (7):815–826
42. Reza N, Hazra LN (2013) Toraldo filters with concentric unequal annuli of fixed phase by evolutionary programming. J Opt Soc Am A 30(2):189–195
43. Hazra LN, Reza N (2010) Superresolution by pupil plane phase filtering. Pramana J Phys 75 (5):855–867
44. de Juana DM, Canales VF, Valle PJ, Cagigal MP (2004) Focusing properties of annular binary phase filters. Opt Commun 229:71–77
45. Sheppard CJR, Calvert G, Wheatland M (1998) Focal distribution for superresolving toraldo filters. J Opt Soc Am A 15(4):849–856
46. de Juana DM, Oti JE, Canales VF, Cagigal MP (2003) Transverse or axial superresolution in a 4Pi-confocal microscope by phase-only filters. J Opt Soc Am A 20(11):2172–2178
47. Yun M, Liu L, Sun J, Liu D (2005) Three-dimensional superresolution by three-zone complex pupil filters. J Opt Soc Am A 22(2):272–277
48. Martinez-Corral M, Andres P, Zapata-Rodriguez CJ, Kowalczyk M (1999) Three-dimensional superresolution by annular binary filters. Opt Commun 165:267–278
49. Ledesma S, Campos J, Escalera JC, Yzuel MJ (2004) Symmetry properties with pupil phase-filters. Opt Express 12(11):2548–2559
50. Canales VF, Oti JE, Cagigal MP (2005) Three dimensional control of focal light intensity distribution by analytically designed phase masks. Opt Commun 247:11–18
51. Cagigal MP, Oti JE, Canales VF, Valle PJ (2004) Analytical design of superresolving phase filters. Opt Commun 241:249–253
52. Piestun R, Shamir J (2002) Synthesis of three dimensional light fields and applications. Proc IEEE 90(2):222–244
53. Konijnenberg AP, Pereira SF (2015) Pupil Engineering to create sheets, lines, and multiple spots at the focal region. J Opt 17(12):125614

54. Straubel CR (1935) "Über Bildgüte", Pieter Zeeman Verhandlungen. Nijhoff, The Hague, pp 302–311
55. Jacquinot P, Roizen-Dossier B (1964) Apodization. In: Wolf E (ed). North Holland, Amsterdam. Prog Opt 3
56. Zernike F (1934) Beugungstheorie des Schneidenver-fahrens und seiner verbesserten form. Physica 1(8):689–704
57. Liu J, Miao E, Sui Y, Yang H (2016) Phase only pupil filter design using Zernike polynomials. J Opt Soc Korea 20(1):101–106
58. Slepian D (1965) Analytic solution of two apodization problems. J Opt Soc Am 55(9):1110–1115
59. Roy Frieden B (1970) The extrapolating pupil, image synthesis, and some thought applications. Appl Opt 9(11):2489–2496
60. Plight M (1978) The rapid calculation of the optical transfer function for on-axis systems using the orthogonal properties of Tchebycheff polynomials. Optica Acta 25(9):849–860
61. Hazra LN, Banerjee A (1976) Application of Walsh function in generation of optimum apodizers. J Opt 5:19–26 (India)
62. Hazra LN (2007) Walsh filters in tailoring of resolution in microscopic imaging. Micron 38 (2):129–135
63. Hazra LN, Guha A (1986) Farfield diffraction properties of radial Walsh filters. J Opt Soc Am A 3(6):843–846
64. De M, Hazra LN (1977) Walsh functions in problems of optical imagery. Opt Acta 24 (3):221–234
65. De M, Hazra LN (1977) On atmospheric turbulence and problem of optimization of the telescopic pupil. Opt Acta 24(3):235–243
66. Mukherjee P, Hazra LN (2013) Farfield diffraction properties of annular Walsh filters. Adv Opt Tech 2013(360450):6
67. Nakamura O, Toyoda K (1991) Side lobe suppression of the point-spread function in annular-pupil optical systems. Appl Opt 30(22):3242–3245
68. Linfoot EH, Wolf E (1953) Diffraction images in systems with an annular aperture. Proc Phys Soc B 66:145–149
69. Sheppard CJR, Wilson T (1979) Imaging properties of annular lenses. Appl Opt 18 (22):3764–3769
70. Boivin A (1964) Théorie et calcul des figures de diffraction de revolution. Gauthier-Villars, Paris
71. Welford WT (1960) Use of annular apertures to increase focal depth. J Opt Soc Am 50:749–753
72. Sheppard CJR, Choudhury A (2004) Annular pupils, radial polarization, and superresolution. Appl Opt 43(22):4322–4327
73. Yun M, Wang M, Liu L (2005) Superresolution with annular binary phase filter in the 4Pi confocal system. J Opt A: Pure Appl Opt 7(11):640–644
74. Mukhopadhyay S, Sarkar S, Bhattacharya K, Hazra LN (2013) Polarization phase shifting interferometric technique for phase calibration of a reflective phase spatial light modulator. Opt Eng 52(3):035602-1–035602-6
75. Ojeda-Castaneda J, Gómez-Reino C (eds) (1996) Selected papers on zone plates. SPIE Optical Engineering Press, Washington
76. Lord Rayleigh III, Experimental Notebook 1870–1878 (U. S. Air Force Geophysics Laboratory Research Library, Hanscom Air Force Base, Massachusetts)
77. Soret JL (1875) Ueber die durch kreisgitter erzeugten Diffractionsphanomene. Ann Phys Chem (Poggendorff) Ser 2 156:99–106
78. Wood RW (1898) Phase-reversal zone-plates and diffraction telescopes. Philos Mag Ser 5 45 (227):511–522

79. Horman MH, Chau HHM (1967) Zone plate theory based on holography. Appl Opt 6
    (2):317–322; Horman MH (1967) Reply to comments on zone plate theory based on
    holography. Appl Opt 6(8):1415–1418; Efficiencies of zone plates and phase zone plates.
    Appl Opt 6(11):2011–2013
80. Dammann H (1970) Blazed synthetic phase-only holograms. Optik 31:95–104 (Stuttgart)
81. Baez AV (1961) Fresnel zone plate for optical image formation using extreme ultraviolet and
    soft x radiation. J Opt Soc Am 51(4):405–412
82. Pfeifer CD, Ferris LD, Yen WM (1973) Optical image formation with a Fresnel zone plate
    using vacuum-ultraviolet radiation. J Opt Soc Am 63(1):91–95
83. Kirz J (1974) Phase zone plates for x rays and the extreme uv. J Opt Soc Am 64(3):301–309
84. Simpson MJ, Michette AG (1984) Considerations of zone plate optics for soft X-ray
    microscopy. Opt Acta 34:1417–1426
85. Tatchyn R, Csonka PL, Landau I (1984) Outline of a variational formulation of zone plate
    theory. J Opt Soc Am B 1(6):806–811
86. Anderson EH (1988) Fabrication technology and applications of zone plates. In: Hoover RB
    (ed) Proceedings of SPIE 1160, X-Ray/EUV optics for astronomy and microscopy, SPIE
    Proceedings, 1990, vol 1160, pp 2–11
87. Michette AG (1986) Optical systems for soft X-Rays. Plenum, New York
88. Anderson EH, Kern D (1992) Nanofabrication of zone plate lenses for X-ray microscopy. In:
    Michette AG, Morrison GR, Buckley CJ (eds) X-ray microscopy III. Springer, Berlin,
    pp 75–78
89. Aristov VV, Basov YA, Snigirev AA (1989) Synchrotron radiation focusing by a
    Bragg-Fresnel lens. Rev Sci Instrum 60:1517–1518
90. Malek CK, Ladan FR, Rivoira R (1991) Fabrication of high-resolution multilayer reflection
    zone plate microlense for the soft X-ray range. Opt Eng 30(8):1081–1085
91. Carnal O, Sigel M, Sleator T, Takuma H, Mlynak J (1991) Imaging and focussing of atoms
    by a Fresnel zone plate. Phys Rev Lett 67:3231–3234
92. Wang S, Zhang X (2002) Terahertz tomographic imaging with a Fresnel lens. Opt Photon
    News 13(12):59
93. Wang Y, Yun W, Jacobsen C (2003) Achromatic Fresnel optics for wideband
    extreme-ultraviolet and X-ray imaging. Nature 424:50–53
94. Kipp L, Skibowski M, Johnson RL, Berndt R, Adelung R, Harm S, Seemann R (2001)
    Sharper images by focusing soft X-rays with photon sieves. Nature 414:184–188
95. Cao Q, Jahns J (2003) Modified Fresnel zone plates that produce sharp Gaussian focal spots.
    J Opt Soc Am A 20(8):1576–1581
96. Cao Q, Jahns J (2004) Comprehensive focusing analysis of various Fresnel zone plates.
    J Opt Soc Am A 21(4):561–571
97. Hazra LN, Han Y, Delisle C (1993) Sigmatic imaging by zone plates. J Opt Soc Am A 10
    (1):69–74
98. Hazra LN, Han Y, Delisle C (1994) Imaging by zone plates: axial stigmatism at a particular
    order. J Opt Soc Am A 11(10):2750–2754
99. Monsoriu JA, Furlan WD, Saavedra G (2005) Focussing light with fractal zone plates.
    Recent Res Devel Opt 5
100. Saavedra G, Furlan WD, Monsoriu JA (2003) Fractal zone plates. Opt Lett 28(12):971–973
101. Zunino L, Garavaglia M (2003) Fraunhofer diffraction by Cantor fractals with variable
     lacunarity. J Mod Opt 50(5):717–727
102. Monsoriu JA, Saavedra G, Furlan WD (2004) Fractal zone plates with variable lacunarity.
     Opt Express 12(18):4227–4234
103. Calatayud A, Ferrando V, Giménez F, Furlan WD, Saavedra G, Monsoriu JA (2013) Fractal
     square zone plates. Opt Commun 286:42–45
104. Ferrando V, Calatayud A, Giménez F, Furlan WD, Monsoriu JA (2013) Cantor dust zone
     plates. Opt Express 21(3):2701–2706
105. Furlan WD, Saavedra G, Monsoriu JA (2007) White-light imaging with fractal zone plates.
     Opt Lett 32(15):2109–2111

106. Mandelbrot BB (1982) The fractal geometry of nature. Freeman, San Francisco
107. Yero OM, Alonso MF, Vega GM, Lancis J, Climent V, Monsoriu JA (2009) Fractal generalised zone plates. J Opt Soc Am A 26(5):1161–1166
108. Gimenez F, Monsoriu JA, Furian WD, Pons A (2006) Fractal photon sieve. Opt Express 14 (25):11958–11963
109. Zhang QQ, Wang JG, Wang MW, Bu J, Zhu SW, Wang R, Gao BZ, Yuan XC (2011) A modified fractal zone plate with extended depth of focus in spectral domain optical coherence tomography. J Opt 13(5):055301 (6 pages)
110. Tao SH, Yuan XC, Lin J, Burge RE (2006) Sequence of focused optical vortices generated by a spiral fractal zone plate. Appl Phys Lett 89(3):031105
111. Monsoriu JA, Furlan WD, Andrés P, Lancis J (2006) Fractal conical lenses. Opt Express 14 (20):9077–9082
112. Melville H, Milne GF (2003) Optical trapping of three-dimensional structures using dynamic holograms. Opt Express 11(26):3562–3567
113. Schonbrun E, Rinzler C, Crozier KB (2008) Microfabricated water immersion zone plate optical tweezer. Appl Phys Lett 92:071112
114. Ashkin A, Dziedzic JM, Bjorkholm JE, Chu S (1986) Observation of a single-beam gradient force optical trap for dielectric particles. Opt Lett 11(5):288–290
115. Neuman KC, Block SM (2004) Optical trapping. Rev Sci Instrum 75(9):2787–2809
116. Dholakia K, Čižmár T (2011) Shaping the future of manipulation. Nat Photonics 5:335–342
117. Molloy JE, Padgett MJ (2002) Lights, action: optical tweezers. Contemp Phys 43(4):241–258
118. Grier DG (2003) A revolution in optical manipulation. Nature 424(6950):810–816 (London)
119. Zhang J, Cao Y, Zheng J (2010) Fibonacci quasi-periodic superstructure fiber Bragg gratings. Optik 121(5):417–421
120. Wu K, Wang GP (2013) One-dimensional Fibonacci grating for far-field super-resolution imaging. Opt Lett 38(12):2032–2034
121. Calatayud A, Ferrando V, Remon L, Furlan WD, Monsoriu JA (2013) Twin axial vortices generated by Fibonacci lenses. Opt Express 21(8):10234–10239
122. Monsoriu JA, Zapata-Rodriguez CJ, Furlan WD (2006) Fractal axicons. Opt Commun 263:1–5
123. Verma R, Banerjee V, Senthilkumaran P (2012) Redundancy in Cantor diffractals. Opt Express 20(8):8250–8255
124. Verma R, Sharma MK, Banerjee V, Senthilkumaran P (2013) Robustness of Cantor diffractals. Opt Express 21(7):7951–7956
125. Gellermann W, Kohmoto M, Sutherland B, Taylor PC (1994) Localization of light waves in Fibonacci dielectric multilayers. Phys Rev Lett 72(5):633–636
126. Yang X, Liu Y, Fu X (1999) Transmission properties of light through the Fibonacci-class multilayers. Phys Rev B 59(7):4545–4548
127. Grushina NV, Korolenko PV, Markova SN (2008) Special features of the diffraction of light on optical Fibonacci gratings. Moscow Univ Phys Bull 63(2):123–126
128. Gao N, Zhang Y, Xie C (2011) Circular Fibonacci gratings. Appl Opt 50(31):G142–G148
129. Verma R, Banerjee V, Senthilkumaran P (2014) Fractal signatures in the aperiodic Fibonacci grating. Opt Lett 39(9):2557–2560
130. Verma R, Sharma MK, Senthilkumaran P, Banerjee V (2014) Analysis of Fibonacci gratings and their diffraction patterns. J Opt Soc Am A 31(7):1473–1480
131. Ferrando V, Gimenez F, Furlan WD, Monsoriu JA (2015) Bifractal focussing and imaging properties of Thue-Morse zone plates. Opt Express 23(15):19846–19853
132. Zhang J (2015) Three-dimensional array diffraction-limited foci from Greek ladders to generalized Fibonacci sequences. Opt Express 23(23):30308–30317
133. Zhang J, Ke J, Zhu J, Lin Z (2015) Three-dimensional array foci of generalized Fibonacci photon sieves. Cornell University. arXive:1510.03511[physics.optics]
134. Mukherjee P, Hazra LN (2014) Self-similarity in radial Walsh filters and axial intensity distribution in the farfield diffraction pattern. J Opt Soc Am A 31(2):379–387

135. Mukherjee P, Hazra LN (2014) Self-similarity in the farfield diffraction patterns of annular Walsh filters. Asian J Phys 23(4):543–560
136. Mukherjee P, Hazra LN (2014) Self-similarity in transverse intensity distributions in the farfield diffraction pattern of radial Walsh filters. Adv Opt 2014(352316):7
137. Born M, Wolf E (1980) Principles of optics. Pergamon, Oxford
138. Goodman JW (1996) Introduction to fourier optics, 2nd edn. McGraw-Hill, Singapore
139. Hopkins HH (1983) Canonical and real space coordinates used in the theory of image formation. In: Shannon RR, Wyant JC (eds) Applied optics and optical engineering. Academic, New York, 9, 307
140. Gu M (2000) Advanced optical imaging theory. Springer, Berlin, pp 46–47
141. Hopkins HH (1981) Calculation of the aberrations and image assessment for a general optical system. Opt Acta 28(5):667–714

Chapter 4
Self-similarity in Transverse Intensity Distributions on the Farfield Plane of Self-similar Walsh Filters

Abstract The set of radial and annular Walsh filters can be classified into distinct self-similar groups and subgroups, where members of each subgroup possess self-similar structures or phase sequences. It has been observed that, the transverse intensity distributions in the farfield diffraction pattern of these self-similar radial and annular Walsh filters are also self-similar. In this chapter we report results of our investigations on the self-similarity in the intensity distributions on a transverse plane in the farfield diffraction patterns of the self-similar radial and annular Walsh filters.

Keywords Self-similar farfield patterns · Self-similar Walsh filters

4.1 Radial Walsh Filters

The radial Walsh filter is placed on the exit pupil of an axially symmetric imaging system (Fig. 4.1) and the normalized intensity distribution on a transverse plane in the farfield of radial Walsh filters is determined by Eq. (3.12).

A point on the transverse image plane in the farfield is represented by the Cartesian co-ordinate (ξ, η) with O' as the origin of the (ξ, η) axes as in Fig. 4.1. The normalized transverse intensity $I_N(p)$ on the farfield diffraction pattern is computed by Eq. (3.12) where p is the reduced diffraction variable or the reduced transverse distance as in Eq. (3.2). Normalized transverse intensity distribution $I_N(p)$ are evaluated for all orders of radial Walsh filters from 0 to 15. Self-similarity in the transverse intensity distributions in the farfield diffraction patterns is observed for each of the self-similar groups of radial Walsh filters [1]. Illustrative results on the same are presented in the following Figs. 4.3, 4.4 and 4.5. Emphasis is given on highlighting the distinctive features observed in the transverse intensity distribution patterns of each self-similar group or subgroup of the radial Walsh filters.

For a Walsh filter of zero order on the exit pupil, the transverse intensity distribution on the farfield plane is the Airy pattern where the normalized central

© The Author(s) 2018

L. Hazra and P. Mukherjee, *Self-similarity in Walsh Functions and in the Farfield Diffraction Patterns of Radial Walsh Filters*, SpringerBriefs in Applied Sciences and Technology, DOI 10.1007/978-981-10-2809-0_4

Fig. 4.1 Farfield diffraction
pattern on the image plane at
O'. Radial Walsh filter $\varphi_\ell(r)$
on the exit pupil

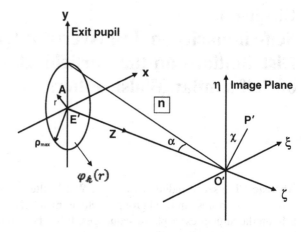

Fig. 4.2 The Airy pattern

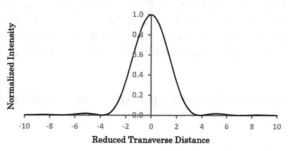

intensity at the origin has magnitude one as in Fig. 4.2. In presence of a Walsh filter
of orders 1, 2, 3... on the exit pupil, the central intensity is zero. The Airy pattern is
also shown in Fig. 4.3, superimposed on the intensity distribution pattern of Walsh
order 1. There is no central lobe or central peak in the transverse intensity distribution on the farfield of any radial Walsh filter, except for order zero. On the
farfield plane, the intensity at the centre is zero, surrounded by bright and dark rings
for all non-zero orders of radial Walsh filters as in Figs. 4.3, 4.4 and 4.5.

Figure 4.3 shows that radial Walsh filters of group I, i.e. orders 1, 3, 7 and 15
produce self-similar transverse intensity distributions on the far field plane. The
intensity at the origin is zero for all the members of this group. This is a common
feature observed in all other groups. Orders 1 and 3 exhibit a distinct maximum just
beside the origin along the transverse direction and fainter side lobes, as illustrated
in Fig. 4.3. The side lobes consist of two sub peaks of almost equal intensity,
prominently noticed in orders 3 and 7. The maximum intensity of the first bright
ring is much larger compared to that of the side lobes. For increasing order of the
filters of this group, the maximum intensity of the first bright ring decreases, and
new side lobes keep originating from the first bright ring itself, and their number
increases with increasing order. This is a unique feature of this group, not observed
in any other group.

Fig. 4.3 Transverse intensity distribution on the farfield of radial Walsh filters of group I, orders 1, 3, 7 and 15. Note that Airy pattern corresponds to the transverse intensity distribution of Walsh filter of order zero

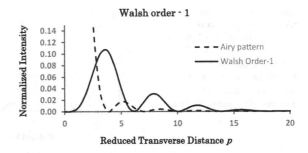

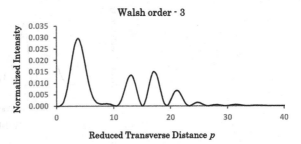

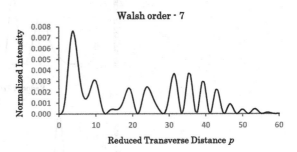

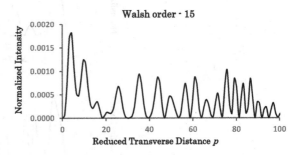

Close observation on this uniqueness reveal that the number of side lobes originating from the first bright ring itself maintain an orderly increment with increase in order number. For order 1, the first bright ring shows no such side lobes and for order 3, the first bright ring is just on the verge of giving rise to a side lobe.

Fig. 4.4 **a** Transverse
intensity distribution on the
farfield of radial Walsh filters
of subgroup IIA, orders 2, 4
and 8. **b** Transverse intensity
distribution on the farfield of
radial Walsh filters of
subgroup IIB, orders 6 and 12

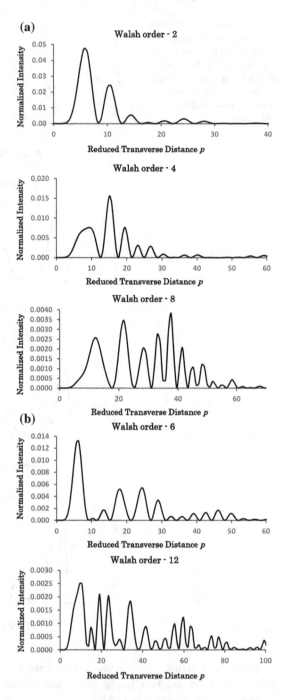

Fig. 4.5 Transverse intensity
distribution on the farfield of
radial Walsh filters of
subgroup IIIAA, orders 5 and
11

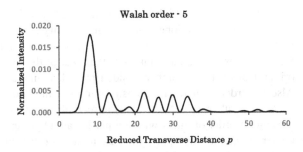

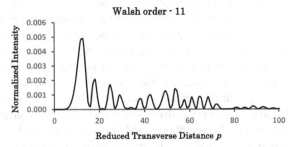

For order 7, one such side lobe is seen and for order 15, it is two. It is also
noteworthy that the minimum intensity of these side lobes do not reach zero. With
increasing order of these filters in the group, the value of these minima keep
decreasing. The number of fainter side lobes also increases, and their intensity
decreases with increasing order. However, the maximum intensity of the first bright
ring and the maximum intensity in the side lobes originating from the first bright
ring have an intensity level much higher than the fainter side lobes. The location of
the peak in the first bright ring beside the origin remain unchanged for orders 1, 3, 7
and then shifts slightly away from the origin for order 15. The horizontal change in
scale in the transverse intensity distributions of the various members in Fig. 4.3 was
made to accommodate the central lobe of highest intensity and few significant side
lobes in the figure.

Subgroup IIA consists of radial Walsh filters of orders 2, 4 and 8. Order 2 exhibit
transverse intensity distribution with a maximum intensity peak just beside the
origin and adjacent fainter side lobes as illustrated in Fig. 4.4a.

The intensity distribution of order 1 of group I and order 2 of subgroup IIA have
almost similar nature with the difference that for subgroup IIA, the intensity of the
first bright ring is lesser and location of this peak is farther from the origin as
compared to the intensity distribution of order 1. The other orders 4 and 8 show
significantly different intensity distribution patterns as compared to group I. Unlike
group I, where the first bright ring maintains a higher intensity level compared to
the side lobes for all increasing orders within the group, in subgroup IIA, as we

move from lower to higher orders, the peak just beside the origin decrease in intensity and the side lobes shoot up in intensity to comparable scale as that of the first bright lobe just beside the origin. For higher order 16 of the subgroup, one of the side lobes have greater intensity than that of the first bright lobe just beside the origin. However, the number of side ripples keep increasing with increase in order. Also for order 8, there are intensity minima which do not reach zero.

As with subgroup IIA, the general trend observed for each subgroup is that, with increase in order of filters within a particular subgroup, as we move from lower to higher orders the side ripples increase in number, the intensity of the lobes near the origin and side ripples fall in magnitude. The decrease in intensity for the first bright lobe near the origin is more compared to the side lobes with the consequence that the side lobes shoot up in intensity to match the scale of the first bright lobe of that particular order. However, each subgroup has a unique self-similar pattern exhibited by its members. Also the nature of the side lobes plays a significant role in distinguishing the intensity distribution of one subgroup from the other.

Subgroup IIB contain Walsh orders 6 and 12. These filters produce self-similar transverse intensity distributions with a maximum intensity peak placed just beside the origin. The neighboring side lobe has three peaks which is a unique feature of this subgroup, followed by ripples as illustrated in Fig. 4.4b. With increasing order of the filters, the transverse intensity patterns follow the general trend. For order 12 there are certain intensity minima which have very low intensity, although not exactly zero.

Members of subgroup IIIAA i.e. radial Walsh filters of orders 5 and 11 produce transverse intensity distribution curves as illustrated in Fig. 4.5. The pattern for order 5 includes a first bright lobe of maximum intensity, placed just beside the origin followed by fainter side lobes of almost equal intensity which is unique and still fainter ripples.

It may be noted in Figs. 4.3, 4.4 and 4.5 that within each subgroup the intensity profiles are self-similar which is a consequence of the inherent self-similarity in members of that subgroup. It may also be noted from these figures that with increasing member order within a subgroup, the lobes become narrower with decreasing intensity. The location of the first bright ring beside the origin, shifts away from the origin with increasing order number of the filters within each subgroup.

Some studies were carried out on other subgroups whose first member falls within the range (0, 15) of ℓ. Subgroup IIIBA order 9, subgroup IVAAA contain Walsh filter of order 10, subgroup IIIAB contain Walsh order 13, subgroup IIC contain Walsh filter of order 14. Filters in each subgroup produce self-similar transverse intensity distributions.

4.2 Annular Walsh Filters

With annular Walsh filters $\varphi_{\ell}^{\varepsilon}(r)$ placed on the exit pupil of an axially symmetric imaging system the farfield intensity distribution is obtained on the transverse image plane. The intensity distribution $I_N^{\varepsilon}(p)$ is expressed by Eq. (3.33). Transverse intensity distribution curves have been plotted for all orders of annular Walsh filters from 0 to 15. Self-similarity is observed in the intensity distributions of the self-similar groups of annular Walsh filters [2]. Discussion on the same is presented in this section.

The set of annular Walsh filters can be classified into distinct self-similar groups and subgroups in the same manner as radial Walsh filters. The transverse intensity distribution produced in the farfield of the self-similar groups of annular Walsh filters are also self-similar which is a consequence of the structural self-similarity of the filters present in a group. Our study is mostly limited to orders ℓ within the range (0, 15). Numerical results have been computed for groups I, IIA, IIB, IIC, IIIAA, IIIAB, IIIBA and IVAAA. This section presents illustrative results depicting self-similarity in the intensity distributions on a transverse plane in the farfield diffraction patterns of the self-similar groups of annular Walsh filters whose first two members fall within the range (0, 15) of, with central obscuration ratio 0.5.

For the transverse intensity distribution by the self-similar annular Walsh filters, the nature of the transverse intensity distribution for a particular order of the filter vary significantly from that of the unobscured pupil. The nature of the pattern changes noticeably with a decrease in the overall intensity in the pattern with increase in value of the central obscuration ratio. However self-similarity is maintained in the transverse intensity distribution of these self-similar groups of annular Walsh filters with a fixed value of the central obscuration ratio ε.

Figures 4.6, 4.7 and 4.8 show the transverse intensity distribution of the self-similar groups of annular Walsh filters with central obscuration ratio $\varepsilon = 0.5$. Figure 4.9 show the transverse intensity distribution for a particular order 3 of the annular Walsh filter, for different values of the central obscuration ratio $\varepsilon = 0.3, 0.5, 0.7$ and 0.9.

It has been noted earlier that the transverse distribution of amplitude or intensity on the farfield plane is altered significantly for increasing order of radial Walsh filters. For annular Walsh filters, it is observed that for filters with fixed value of central obscuration ratio ε, self-similarity is maintained in the transverse intensity distributions in the farfield for a specific group of self-similar annular Walsh filters.

The general trend observed in the transverse intensity pattern of the annular Walsh filters is that the central intensity is always zero for all orders of the filter, except the zeroth order that gives rise to maximum intensity at the centre of the diffraction pattern. The dark centre is surrounded by rings of oscillating intensity. For lower orders, this oscillation gradually decays after the first few rings, but for higher orders, the oscillation continue for many more rings. As expected from energy considerations, the oscillation is significantly less in case of lower order filters compared to the higher order ones in each case of obscuration. As we move from lower to higher obscuration, the ring with peak intensity gradually shifts away from the origin.

Fig. 4.6 Transverse intensity distribution on the farfield of annular Walsh filters of group I, orders 1, 3, 7 and 15

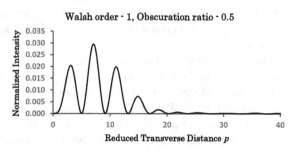

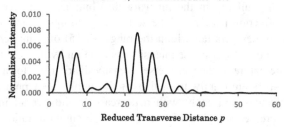

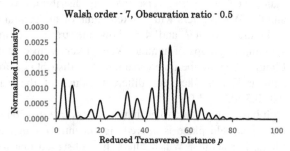

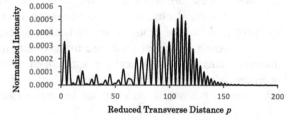

Group I members i.e. annular Walsh filters of orders 1, 3, 7 and 15 with obscuration ratio 0.5 produce self-similar transverse distributions as in Fig. 4.6. For order 1, three peaks are seen just beside the origin followed by fainter side ripples. Of the three peaks, the central peak has highest intensity and the envelope to these peaks is almost a sinusoid. With increasing order of the members of this group, this peak intensity shifts away from the origin and a dual peak emerges just beside the origin. These dual peaks have lesser intensity than the triple peaks. For order 3, there is a ripple with two faint peaks between the dual and triple peaks. For increasing orders 7 and 15, the dual and triple peaks become narrower, fall in intensity and more ripples originate between them. Looking closely at the dual peak, the one closer to the origin, has higher intensity and if we consider the faint ripples between the dual and the triple peaks, they are also double peaked with the higher intensity lobe placed away from the origin for each pair of such ripples formed.

Subgroup IIA contain annular Walsh filters of order 2, 4 and 8. For order 2, the transverse intensity distribution shows three peaks with two faint sub peaks on either side of it, placed near the origin and the envelope to these peaks is a sinusoid as in Fig. 4.7a. Of these peaks the central peak has highest intensity. With increase in order number within the subgroup, the overall intensity of the pattern decreases, the peak intensity shifts away from the origin and subpeaks of lesser intensity and side ripples keep originating. The sub-peaks also shoot up in intensity with increasing order of the filter. Between the two pairs, lobes of almost equal intensity which have intensity comparable to the two maxima are visible. The envelope to these peaks is almost a sinusoid with a dip in its peak.

Subgroup IIB contains Walsh orders 6 and 12. The intensity distribution of order 6 show a pair of dual peaks away from the origin followed by fainter ripples as in Fig. 4.7b. The dual peaks themselves have their associated side lobes and the envelope to each of these two clusters of peaks is almost sinusoidal. With increasing order within the subgroup, the overall intensity of the pattern decreases, the lobes become narrower and each dual peak gives rise to additional side lobes. These side lobes have intensity somewhat comparable with that of the peak intensity. A consequence is that the two clusters lose their dual peaked nature and display multiple peaks. The cluster of peaks closer to the origin display higher intensity than the cluster away from it. The envelope to the second cluster of peaks from the origin is no more a single sinusoid as in order 6, but consists of two sinusoidal humps in order 12.

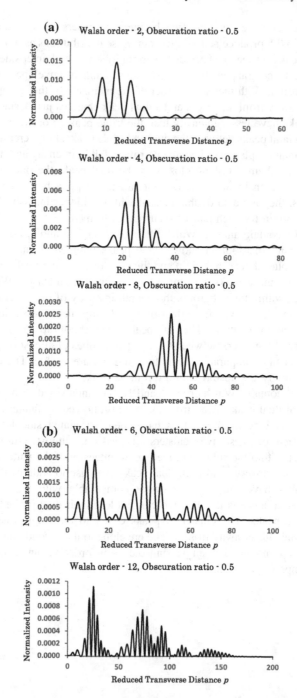

Fig. 4.7 **a** Transverse intensity distribution on the farfield of annular Walsh filters of subgroup IIA, orders 2, 4 and 8. **b** Transverse intensity distribution on the farfield of annular Walsh filters of subgroup IIB, orders 6 and 12

Fig. 4.8 Transverse intensity
distribution on the farfield of
annular Walsh filters of
subgroup IIIAA, orders
5 and 11

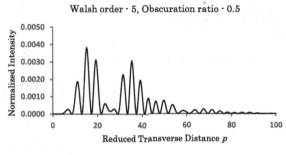

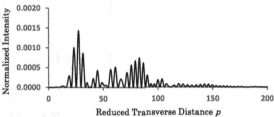

Subgroup IIIAA contain Walsh orders 5 and 11. The transverse intensity distribution of order 5, as in Fig. 4.8 show a pair of triple peaks placed near the origin. Each triple peak has its associated side lobes and the envelope to these two peaks are almost sinusoidal humps. The two humps having the range of peaks are separated by small ripples. The hump closer to the origin has the peak with highest intensity. With increasing order of the members in the group, there is an overall decrease in intensity in the pattern, each triple peak gives rise to sub peaks, and multiple peaks emerge. Of the two humps in order 5, the hump closer to the origin with three peaks give rise to multiple peaks for higher orders, but the envelope maintains the single sinusoidal hump. For the second hump away from the origin, the decrease in intensity of the range of peaks is more prominent and the envelope to these range of peaks contain many small sinusoidal humps of varying spatial frequency.

Figure 4.9 shows the transverse intensity distribution for annular Walsh filters of order 3, for various values of the central obscuration ratio $\varepsilon = 0.3, 0.5, 0.7$ and 0.9. For a particular order, with increasing value of the obscuration ratio ε, the nature of the intensity distribution becomes more pronounced with the frequency of oscillation in intensity becoming increasingly higher with concomitant decrease in peak intensity. With increasing obscuration, the envelope to the distribution show more defined sinusoidal humps as compared to that with lower obscuration values and the peak intensity shifts further away from the origin. For lower values of the obscuration ratio, there are some intensity minima which do not return to zero as seen for $\varepsilon = 0.3$ and 0.5. For higher values of the obscuration ratio $\varepsilon = 0.7$ and 0.9 all the intensity minima reach zero.

For other orders of the annular Walsh filters, observations are mostly similar.

Fig. 4.9 Effects of change in central obscuration ratio ε of the annular Walsh filter $\varphi_3^\varepsilon(r)$ on the transverse intensity distribution on the farfield

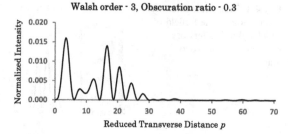

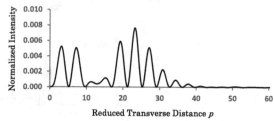

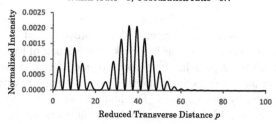

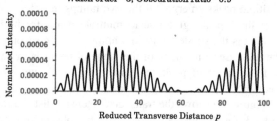

References

1. Mukherjee P, Hazra LN (2014) Self-similarity in transverse intensity distributions in the farfield diffraction pattern of radial Walsh filters. Adv Opt 2014:7 (Article ID 352316)
2. Mukherjee P, Hazra LN (2014) Self-similarity in the farfield diffraction patterns of annular Walsh filters. Asian J Phys 23(4):543–560

Chapter 5
Self-similarity in Axial Intensity Distributions of Self-similar Walsh Filters

Abstract The set of radial and annular Walsh filters can be classified into distinct self-similar groups and subgroups, where members of each subgroup possess self-similar structures or phase sequences. It has been observed that the axial intensity distributions around the focal/image plane are also self-similar when these self-similar radial and annular Walsh filters are used as pupil plane filters. In this chapter we report results of our investigations on the self-similarity in the intensity distributions around the focal/image plane when self-similar radial and annular Walsh filters are used as pupil filters.

Keywords Self-similar axial intensity · Self-similar Walsh filters

5.1 Characteristics of Axial Intensity Distributions Around the Image Plane with Radial Walsh Filters on the Exit Pupil

Figure 5.1 shows the image space of an axially symmetric imaging system. The exit pupil is located at E', and the image plane is at O' on the axis. With a radial Walsh filter $\varphi_k(r)$ placed on the exit pupil, the normalized intensity at an axial point O'' that is axially shifted from the image plane is computed by using Eq. (3.24). The plane where O'' is located is shifted longitudinally from the image/focal plane by $\Delta\zeta$. a represents the reduced axial co-ordinate or distance as expressed by Eq. (3.16). Normalized intensities have been computed for different values of reduced axial distance a. Axial intensity distribution curves have been plotted for all orders of radial Walsh filters from 0 to 15, and are presented. The origin of the axial intensity distribution curves represent the location of the image/focal plane (Fig. 5.1) and the positive and negative values of a represent axial shifts on either side of this plane. Self-similarity has been observed in the axial intensity distributions of the

© The Author(s) 2018

L. Hazra and P. Mukherjee, *Self-similarity in Walsh Functions and in the Farfield Diffraction Patterns of Radial Walsh Filters*, SpringerBriefs in Applied Sciences and Technology, DOI 10.1007/978-981-10-2809-0_5

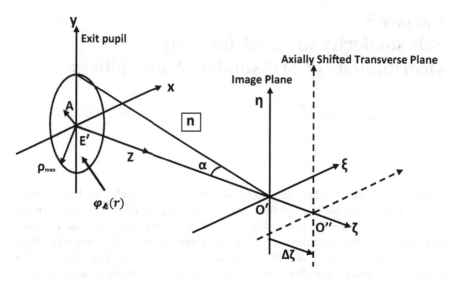

Fig. 5.1 Radial Walsh filter $\varphi_k(r)$ on the exit pupil of an axisymmetric imaging system

self-similar groups of radial Walsh filters [1]. Details of such observations are presented in what follows. Emphasis is given on highlighting the distinctive features in the axial intensity distribution patterns of each self-similar group or subgroup of radial Walsh filters.

Placing a radial Walsh filter of order zero, i.e. $\varphi_0(r)$ on the exit pupil does not alter the intensity distribution, since $\varphi_0(r)$ does effectively represent an Airy pupil, by definition. The corresponding axial intensity distribution around the image/focal plane is shown in Fig. 5.2. The axial intensity is maximum on the image/focal

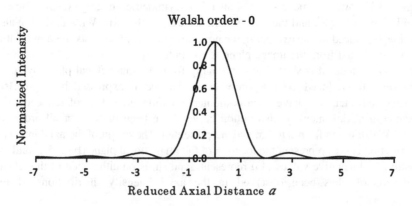

Fig. 5.2 Distribution of axial intensity around the image/focal plane with a radial Walsh filter of order zero on the exit pupil

plane, and the intensity distribution along the axis is symmetric around it. For all radial Walsh filters of integral order other than zero, the axial intensity on the image/focal plane is zero, and the intensity along the axis is same for both $+\Delta\zeta$ and $-\Delta\zeta$, where $\Delta\zeta$ represents the distance of the point O'' from O' as in Fig. 5.1.

On the basis of the classification made in Chap. 2, axial intensity distribution curves around the image plane are computed for the first few elements of some of the groups of self-similar Walsh filters. The first four members of group I of radial Walsh filters are the filters $\varphi_1(r)$, $\varphi_3(r)$, $\varphi_7(r)$ and $\varphi_{15}(r)$. The axial intensity distributions around the image plane obtained with these filters on the exit pupil are presented in Fig. 5.3. The axial intensity distribution of this group of filters consists of two symmetric patterns along the axis on both sides of the point O'. Two major lobes of equal intensity characterize this pattern. The distance between these two lobes gradually increases with increase in order of the radial Walsh filters of the same group, and the lobes become increasingly narrower with the peak intensity remaining almost unchanged. Increase in order of the filters of group I also leads to a gradual increase in the number of the minor lobes in the axial intensity pattern, but the intensity of the minor lobes decreases sharply. Note the changes in scale along the abscissa in one of the axial intensity distribution curves of Fig. 5.3. This was necessary to accommodate the two major lobes in the figure.

Figure 5.4a presents the distribution of axial intensity around the image plane with radial Walsh filters belonging to subgroup IIA on the exit pupil of the imaging system. The first three members of this subgroup are the filters $\varphi_2(r)$, $\varphi_4(r)$ and $\varphi_8(r)$. The axial intensity distributions are characterized by dual maxima lying on both sides of the image plane, and faint side lobes. Unlike group I members, where ripples originate near the origin between the two primary lobes, for lower order members of subgroup IIA, the side lobes are placed away from the origin. On both sides of the image plane, the two primary maxima are placed near the origin with no ripple between them. Radial Walsh filters of orders 6 and 12, i.e. $\varphi_6(r)$ and $\varphi_{12}(r)$ are the first two members of the subgroup IIB. The intensity distribution curves of Fig. 5.4b show that these filters produce self-similar axial intensity distributions with pairs of triple maxima located symmetrically around the image plane. For each of the triplets, the central lobe has highest intensity. The secondary maximum with higher intensity is placed away from the origin. For radial Walsh filter of order 12, ripples are discernible between the primary maximum and the secondary maximum placed away from the origin.

Radial Walsh filters, $\varphi_5(r)$ and $\varphi_{11}(r)$ are the first two members of subgroup IIIAA and Fig. 5.5 shows that the axial intensity distributions produced by members of this subgroup are characterized by a pair of triple maxima, located symmetrically on both sides of the image plane. Alike members of the earlier group, the central lobe has highest intensity in each of the triplets. However, the secondary maximum with higher intensity is placed towards the origin in case of the filter $\varphi_5(r)$. For filter $\varphi_{11}(r)$, ripples are discernible between the primary maximum and the secondary maximum placed near the origin.

Fig. 5.3 Distribution of axial intensity around the image/focal plane with first few radial Walsh filters of group I on the exit pupil

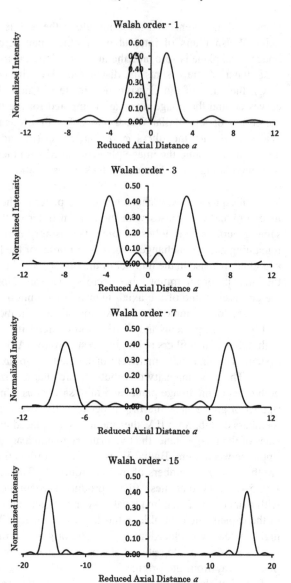

Results presented in Figs. 5.3, 5.4 and 5.5 validates the conjecture that the distribution of axial intensity around the image/focal plane, with self-similar radial Walsh filters on the exit pupil of an imaging system, are self-similar. It is also significant to note that with increase in order k of members within a subgroup, the main lobes become narrower with increase in distance between the lobes. In general, this is accompanied by an increase in the number of ripples, albeit of decreasing intensity.

Fig. 5.4 **a** Distribution of axial intensity around the image/focal plane with first few radial Walsh filters of subgroup IIA. **b** Distribution of axial intensity around the image/focal plane with first few radial Walsh filters of subgroup IIB

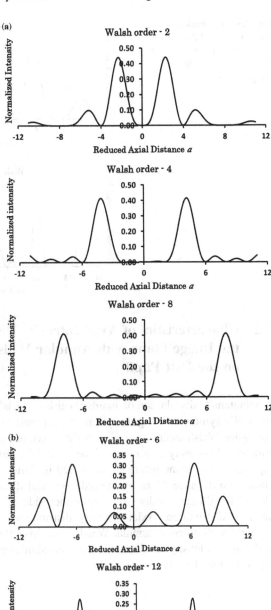

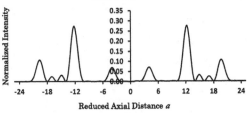

Fig. 5.5 Distribution of axial intensity around the image/focal plane with first few radial Walsh filters of subgroup IIIAA

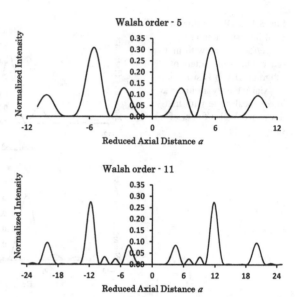

5.2   Characteristics of Axial Intensity Distribution Around the Image Plane with Annular Walsh Filters on the Exit Pupil

Axial intensity distribution of annular Walsh filters $\varphi_k^\varepsilon(r)$ placed on the exit pupil of an axially symmetric imaging system are determined in the same manner as that of the radial Walsh filters discussed in the previous section. The expression for the normalized intensity at a point O'' on an axially shifted image plane is given by Eq. (3.34). The plane where O'' is located is shifted longitudinally from the image plane by a distance $\Delta\zeta$ or by the reduced axial distance a. Normalized intensities plotted for different values of a on either side of the origin O' gives the axial intensity distribution. Axial intensity distribution curves have been plotted and studied for all orders of annular Walsh filters from 0 to 15. Self-similarity observed in the axial intensity distribution of the self-similar groups of annular Walsh filters [2] is discussed in the next section.

Fig. 5.6 Distribution of axial
intensity around the
image/focal plane with first
few annular Walsh filters of
group I; obscuration ratio: 0.3

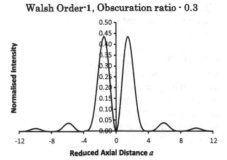

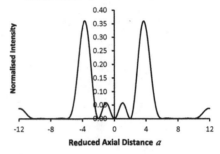

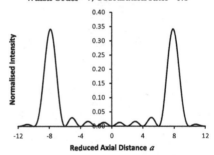

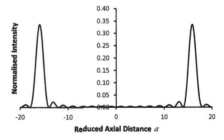

Fig. 5.7 a Distribution of
axial intensity around the
image/focal plane with first
few annular Walsh filters of
subgroup IIA; obscuration
ratio: 0.3. **b** Distribution of
axial intensity around the
image/focal plane with first
few annular Walsh filters of
subgroup IIB; obscuration
ratio: 0.3

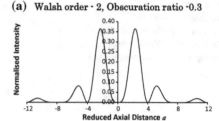

(a) Walsh order ‑ 2, Obscuration ratio ‑0.3

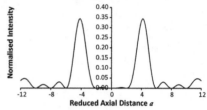

Walsh Order ‑ 4, Obscuration ratio ‑ 0.3

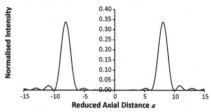

Walsh Order ‑ 8, Obscuration ratio ‑ 0.3

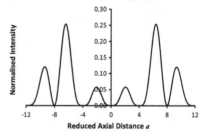

(b) Walsh Order ‑ 6, Obscuration ratio ‑0.3

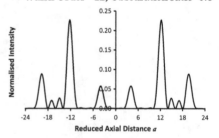

Walsh Order ‑ 12, Obscuration ratio ‑0.3

Our study is mostly limited to orders ℓ within the range (0, 15) with occasional foray in annular Walsh functions of neighboring higher orders. Numerical results have been computed for groups I, IIA, IIB, IIC, IIIAA, IIIAB, IIIBA and IVAAA. As an illustrative example, all results presented below correspond to annular filters with obscuration ratio 0.3 and 0.5. Self-similarity has been observed in the axial intensity distributions of self-similar groups of annular Walsh filters. The nature of the axial intensity distribution for a particular order of the annular Walsh filter, remain the same as that produced by an unobscured pupil i.e. the radial Walsh filter of the same order, for all values of obscuration ratio. It should be noted however, that for a particular order of the annular Walsh filter, as we increase the obscuration ratio the magnitude of intensity in the axial distribution decreases.

Figures 5.6, 5.7, 5.8, 5.9, 5.10 and 5.11 shows the self-similarity in the axial intensity distributions around the image plane when members of self-similar groups of annular Walsh filters are used as pupil filters. Figures 5.6, 5.7 and 5.8 deal with annular Walsh filters with obscuration ratio 0.3, and Figs. 5.9, 5.10 and 5.11 correspond to annular filters with obscuration ratio 0.5. Figure 5.12 shows the decrease in intensity in the axial distribution with increase in obscuration ratio for an annular Walsh filter of a particular order. It should be noted that both the shape of the individual lobes and the axial distance between the two major lobes remains same with change in obscuration ratio of any particular annular Walsh filter. However, the peak intensity of the lobes decreases with increase in obscuration ratio.

Fig. 5.8 Distribution of axial intensity around the image/focal plane with first few annular Walsh filters of subgroup IIIAA; obscuration ratio: 0.3

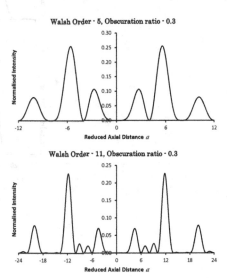

Fig. 5.9 Distribution of axial intensity around the image/focal plane with first few annular Walsh filters of group I; obscuration ratio: 0.5

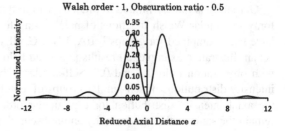

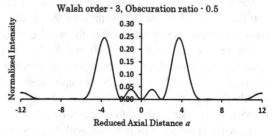

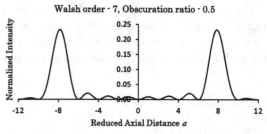

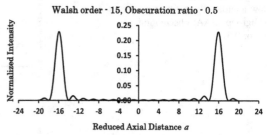

Fig. 5.10 a Distribution of axial intensity around the image/focal plane with first few annular Walsh filters of subgroup IIA; obscuration ratio: 0.5. **b** Distribution of axial intensity around the image/focal plane with first few annular Walsh filters of subgroup IIB; obscuration ratio: 0.5

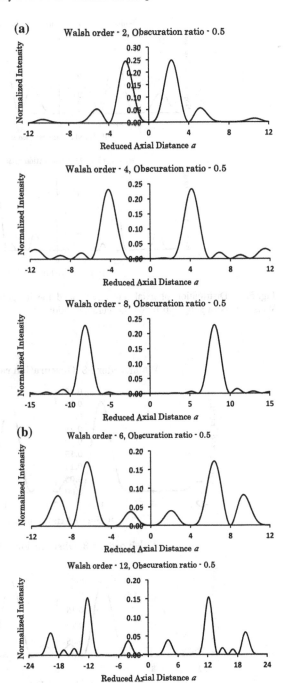

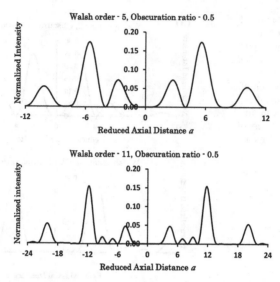

Fig. 5.11 Distribution of axial intensity around the image/focal plane with first few annular Walsh filters of subgroup IIIAA; obscuration ratio: 0.5

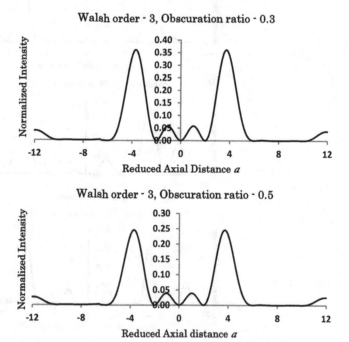

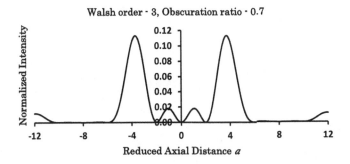

Walsh order - 3, Obscuration ratio - 0.7

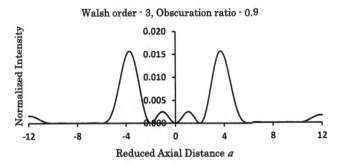

Walsh order - 3, Obscuration ratio - 0.9

Fig. 5.12 Effects of change in central obscuration ratio ε of the annular Walsh filter $\varphi_3^\varepsilon(r)$ on the distribution of axial intensity around the image/focal plane

References

1. Mukherjee P, Hazra LN (2014) Self-similarity in radial Walsh filters and axial intensity distribution in the farfield diffraction pattern. J Opt Soc Am A 31(2):379–387
2. Mukherjee P, Hazra LN (2014) Self-similarity in the farfield diffraction patterns of annular Walsh filters. Asian J Phys 23(4):543–560

Chapter 6
Self-similarity in 3D Light Distributions Near the Focus of Self-similar Radial Walsh Filters

Abstract The three dimensional light distribution near the focus of rotationally symmetric imaging system may be tailored by the technique of pupil plane filtering using Walsh filters. The set of Walsh filters can be classified into distinct self-similar groups and subgroups, where members of each subgroup possess self-similar structures or phase sequences. The 3D light distribution around the focal/image plane portrays self-similarity which can be correlated to the structural self-similarity of Walsh filters themselves. In this chapter we report results of our investigations on the self-similarity in the three dimensional intensity distribution near the focus of the self-similar radial Walsh filters.

Keywords Self-similar optical traps · Self-similar Walsh filters · Self-similar optical tweezers

6.1 Introduction

Pupil plane filtering may be used to reshape the spatial intensity distribution near the focus to cater to apodization or superresolution [1, 2]. Annular pupil-plane masks are used to control light intensity near the focus and find applications in microscopy and optical tweezers [3]. Synthesis of three dimensional light fields have also been done using micro structures on diffractive optical elements [4] which could find application in lithography where depth of focus is as important as resolution, and in volume memories or near field optics. Shamir et al. [5–7] used diffractive optical elements to create unconventional three dimensional light distributions or sculptured the three dimensional light fields with potential for applications in instrumentation, optical computing, optical interconnections and material processing. Dark beams may be used for atom trapping and micro particle handling applications. An adjustable three dimensional dark focus was also generated surrounded by light in all three dimensions [8]. It finds applications such as optical tweezers and atom trapping.

The orthogonality and self-similarity of the radial Walsh filters can be harnessed to sculpture the 3D light distribution near the focus for many potential applications in the field of optical micromanipulation.

© The Author(s) 2018   73
L. Hazra and P. Mukherjee, *Self-similarity in Walsh Functions and in the Farfield Diffraction Patterns of Radial Walsh Filters*, SpringerBriefs in Applied Sciences and Technology, DOI 10.1007/978-981-10-2809-0_6

With this goal in view, three dimensional light distribution near the focus of an axially symmetric imaging system with a radial Walsh filter placed on the exit pupil is studied in this chapter. Normalized transverse intensity distribution i.e. variation of normalized intensity with reduced transverse distance p is computed for various transverse planes that are longitudinally shifted from the paraxial focus O' by longitudinal distance $\Delta\zeta$ or reduced axial distance a. Mathematical formulation for computing the intensity distributions is presented in the following Sect. 6.2. The 3D light distributions near the focus have been studied for all orders of radial Walsh filters from 0 to 15. Self-similarity has been observed in the 3D light distributions of the self-similar groups of radial Walsh filters. Section 6.3 presents illustrations which show the cross-sections of the intensity distributions along the (χ, ζ) plane in the farfield of radial Walsh filters of order 0–15. Self-similarity observed in these light distributions of the self-similar groups of radial Walsh filters have also been discussed.

6.2   Computation of Intensity Distributions on Transverse Planes in the Focal Region with Radial Walsh Filters on the Exit Pupil

Following the analysis presented earlier in Sects. (3.2) and (3.3) of Chap. 3, the normalized complex farfield amplitude distribution on a transverse plane located at an axial distance of $\Delta\zeta$ from the image/focal plane (Fig. 6.1) is given by

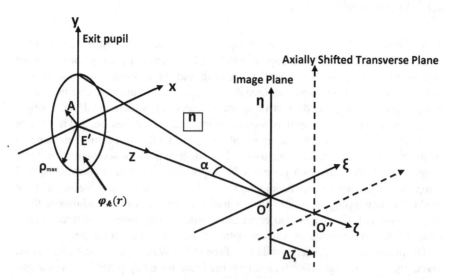

Fig. 6.1 Exit pupil and focal/image plane in the image space of an axially symmetric imaging system. Radial Walsh filter $\varphi_k(r)$ on the exit pupil

$$F_N(p,a) = F_N(\chi, \Delta\zeta) = 2 \sum_{m=1}^{M} \exp[ik\psi_m] \sum_{j=1}^{J} \exp\left[ik\overline{W}_m^j\right] \int_{r_m^{j-1}}^{r_m^j} J_0(pr)\, r dr \quad (6.1)$$

where $\overline{W}_m^j$ is the average value of the wave aberration function over the jth subzone of the mth zone and is given by

$$\overline{W}_m^j = \frac{\int_{r_m^{j-1}}^{r_m^j} W(r)\, r dr}{\int_{r_m^{j-1}}^{r_m^j} r dr} = \frac{W_{20}}{2T} \left[2(m-1)J + (2j-1)\right] \quad (6.2)$$

and

$$\int_{r_m^{j-1}}^{r_m^j} J_0(pr)\, r dr = \mathcal{J}_m^j(p) = \left[\frac{r_m^j J_1\left(pr_m^j\right) - r_m^{j-1} J_1\left(pr_m^{j-1}\right)}{p}\right] \quad (6.3)$$

$J_1(x)$ is the first order Bessel function of argument x. We rewrite below the relations (3.2) and (3.16) that express the relations between the normalized distances p and a with corresponding geometrical distances respectively.

$$p = \frac{2\pi}{\lambda}(n\sin\alpha)\chi \quad (6.4)$$

$$a = \frac{1}{n\lambda}(n\sin\alpha)^2\Delta\zeta \quad (6.5)$$

From (6.1) to (6.3)

$$F_N(p,a) = 2 \sum_{m=1}^{M} \exp[ik\psi_m] \sum_{j=1}^{J} \exp\left[ik\overline{W}_m^j\right] \mathcal{J}_m^j(p) \quad (6.6)$$

The normalized intensity distribution on a transverse plane located at an axial distance of $\Delta\zeta$ from the image/focal plane is given by $I_N(p,a) = |F_N(p,a)|^2$

$$I_N(p,a) = 4 \sum_{m=1}^{M} \sum_{u=1}^{M} \sum_{j=1}^{J} \sum_{v=1}^{J} \cos\left[\{k\psi_m - k\psi_u\} + \left\{k\overline{W}_m^j - k\overline{W}_u^v\right\}\right] \mathcal{J}_m^j(p) \mathcal{J}_u^v(p)$$

$$(6.7)$$

6.3   Illustrative Results

The self-similar 3D light distributions in transverse planes around the focal plane produced by some of the self-similar groups of radial Walsh filters are put forward in this section.

A cross section of the intensity distribution along (χ, ζ) plane in the farfield diffraction pattern for the self-similar groups of radial Walsh filters $\varphi_{\ell}(r)$, $\ell = 0, 1, 2, \ldots, 15$, whose first two members lie within order 15 is shown in Figs. 6.2, 6.3, 6.4 and 6.5.

On account of axial symmetry of the imaging system, the intensity distribution is the same along any plane containing the ζ axis. The elliptical blobs or patches of light seen in the above mentioned figures are the cross sections of the 3D intensity distributions produced near the focus. Except for the case of $\varphi_0(r)$, where a single structure is observed, in all other cases multiple structures are observed. However, the pattern of light distribution or patches produced for each self-similar groups of radial Walsh filters are self-similar, reflecting the inherent self-similarity in the structures of the Walsh filters themselves.

Group I, i.e. radial Walsh filters of orders 1, 3, 7 and 15, produce self-similar three dimensional intensity distributions, with distinct dual maxima on either side of the origin, along the axial direction. The cross-section of this intensity distribution along the (χ, ζ) plane shows distinct dual elliptical patches of light on either side of the origin as illustrated in Fig. 6.3. With increasing order number within the group, the distance between the patches increase and the size of the individual patches decrease. This is also a general trend observed in all other subgroups.

Subgroup IIA consist of radial Walsh filters of orders 2, 4 and 8. They produce three dimensional intensity distribution with a dual maxima placed on either side of

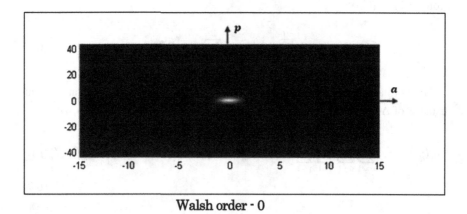

Walsh order - 0

Fig. 6.2 Section of 3D intensity distribution along the (χ, ζ) plane for Walsh order 0

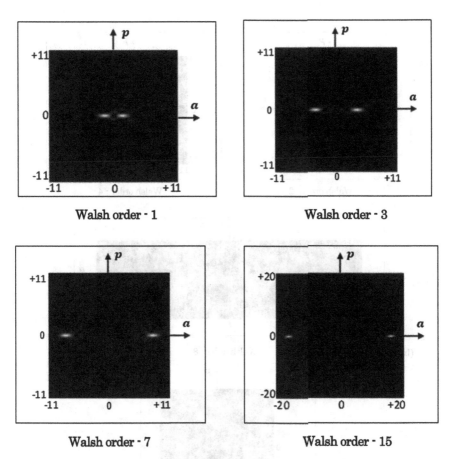

Fig. 6.3 Section of 3D intensity distribution along the (χ, ζ) plane for group I, Walsh orders 1, 3, 7 and 15

the origin and fainter side lobes placed away from the origin. The cross-section of this intensity distribution along the (χ, ζ) plane shows elliptical patches of light on either side of the origin and fainter and smaller patches beside them for the lowest order 2. With increasing order number of members within the subgroup, the patches become smaller in size, the distance between them increases and the side lobes or patches become practically invisible as their intensity decreases as shown in Fig. 6.4a.

Subgroup IIB, i.e. Walsh orders 6 and 12 show 3D intensity distribution with pairs if triple maxima located symmetrically on either side of the origin. For one of the triplets, on one side of the origin, the central lobe has highest intensity and the secondary maximum with higher intensity is placed away from the origin. The cross-sectional view shows a pair of three elliptical patches, each pair located on

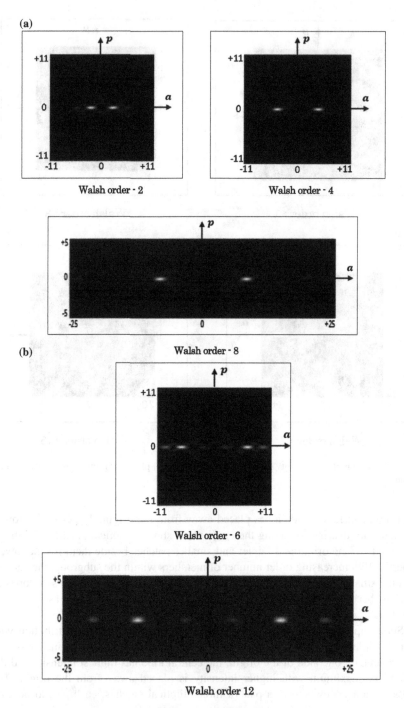

Fig. 6.4 a Section of 3D intensity distribution along the (χ, ζ) plane for subgroup IIA, Walsh orders 2, 4 and 8, **b** section of 3D intensity distribution along the (χ, ζ) plane for subgroup IIB, Walsh orders 6 and 12

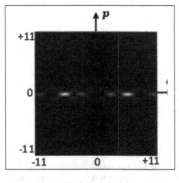

Walsh order - 5

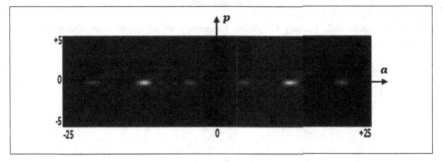

Walsh order - 11

Fig. 6.5 Section of 3D intensity distribution along the (χ, ζ) plane for subgroup IIIAA, Walsh orders 5 and 11

either side of the origin. Considering a triplet on one side of the origin, for order 6, the central patch is the brightest with two fainter patches on either side of it. Of the two patches, the one away from the origin is brighter than the one towards the origin, as illustrated in Fig. 6.4b. This feature is also observed for the higher order 12 with the only difference that the distance between the light patches increase and their size decreases with increase in order number of the members of the subgroup.

Subgroup IIIAAA contain Walsh filters of orders 5 and 11. They produce 3D intensity distributions with pairs of triple maxima located on either side of the origin. Considering a triplet on one side of the origin, the central lobe has highest intensity and the secondary maximum with higher intensity is placed towards the origin for order 5. The cross-sectional view shows a pair of three elliptical patches, each pair located on either side of the origin. Considering a pair on one side of the origin, the central patch is brightest and largest as illustrated in Fig. 6.5. The patch closer to the origin represents the secondary maximum with higher intensity which is fainter than the central patch but is brighter than the patch placed away from the origin. These two patches are smaller in size compared to the central patch. With higher order 11, the self-similarity is maintained with the only difference that the size of the patches and their intensity decrease and they are placed farther apart.

References

1. Ledesma S, Campos J, Escalera JC, Yzuel MJ (2004) Symmetry properties with pupil phase-filters. Opt Exp 12(11):2548–2559
2. Canales VF, Oti JE, Cagigal MP (2005) Three dimensional control of focal light intensity distribution by analytically designed phase masks. Opt Commun 247:11–18
3. Cagigal MP, Oti JE, Canales VF, Valle PJ (2004) Analytical design of superresolving phase filters. Opt Commun 241:249–253
4. Piestun R, Shamir J (2002) Synthesis of three dimensional light fields and applications. Proc IEEE 90(2):222–244
5. Shamir J, Piestun R (2002) Sculpturing of three dimensional light fields. Holography for the new millennium proceedings of SPIE, vol 6483 Springer, Berlin
6. Piestun R, Spektor B, Shamir J (1996) Wave fields in three dimensions: analysis and synthesis. J Opt Soc Am A 13(9):1837–1848
7. Piestun R, Spector B, Shamir J (1996) Unconventional light distributions in three-dimensional domains. J Mod Opt 43(7):1495–1507
8. Yelin D, Bouma BE, Tearney GJ (2004) Generating an adjustable three-dimensional dark focus. Opt Lett 29(7):661–663

Chapter 7
Concluding Remarks

Abstract The investigations reported in this monograph conclusively demonstrate that the three dimensional point spread function of radial and annular Walsh filters portray self-similarity which can be correlated to the self-similar structures of the diffracting apertures. Self-similarity and orthogonality of these filters may be harnessed to solve challenging problems for generation of prespecified 3D patterns. This chapter underscores the importance of the study reported in this monograph for tackling problems of wave optical engineering.

Keywords Self-similarity · Pupil filters · Radial Walsh filters · Annular Walsh filters · Optical tweezers

The radial Walsh filters corresponding to Group I of Walsh functions [vide Table 2.1, Chap. 2] constitute the well-known set of binary phase zone plates. The axial peaks in farfield intensity distribution of the zone plates are interpreted in terms of diffraction orders with specified diffraction efficiencies that can be determined analytically. Radial Walsh filters corresponding to other self-similar groups of Walsh functions, as enunciated earlier in Chap. 2 of this monograph, and different groups of annular Walsh filters in general, constitute variants of binary phase zone plates and display unique diffraction characteristics. The axial peaks in these patterns are located in different positions along the axis; their diffraction efficiencies are also different. It remains to explore analytical methods for determining the location of these axial peaks and also their diffraction efficiencies. Finally, it may be reiterated that Walsh filters, particularly those of higher order open up new frontiers for synthesis of useful phase filters to be utilized in case of high numerical aperture systems. This constitutes new challenges to be tackled by using vectorial diffraction theory.

Three types of pupil filters, namely phase only, amplitude only, and mixed type with variation in both amplitude and phase, are explored to tackle different practical problems in wave optical engineering. From energy considerations the first type is lossless, the second type is lossy, and the third type is called a leaky filter. While the radial Walsh filters fall in the lossless phase only category, the annular Walsh filters are leaky filters of the mixed type. Of course, in matched obscured aperture

© The Author(s) 2018

L. Hazra and P. Mukherjee, *Self-similarity in Walsh Functions and in the Farfield Diffraction Patterns of Radial Walsh Filters*, SpringerBriefs in Applied Sciences and Technology, DOI 10.1007/978-981-10-2809-0_7

systems, the annular Walsh filters act as appropriate lossless filter. Indeed, an annular Walsh filter constitute an interesting mixed filter with total obscuration over a central circular zone surrounded by an annular lossless zone. In general, the obscuration ratio provides a degree of freedom that can be explored to cater to the demands of challenging problems.

Practical implementation of pupil filters based on radial and/or annular Walsh filters is relatively easier compared to pupil filters with continuously varying phase and/or amplitude, for the former filters are piecewise constant, i.e. they consist of a finite number of zones, on each of which the value of phase and/or amplitude is fixed. Indeed, suitable choices on the number, and shape of the zones on the pupil, as well as the discrete number of amplitude and/or phase levels can be incorporated in the analysis from the inception. Currently the ready availability of high efficiency spatial light modulators has greatly facilitated practical realization of these filters in practice [1].

Besides applications in problems of information processing, digital image processing [2–4], optical image formation, apodization, adaptive optics, tailoring of resolution in microscopic imaging and optical encoding [5–9], the feature of self-similarity of radial and annular Walsh filters can be harnessed in practice to produce complex 3D light distributions near the focus to cater to the needs of 3D imaging, lithography, optical superresolution, optical micromanipulation and optical tomography, to mention a few [10]. The two characteristics, namely, the orthogonality of radial and annular Walsh filters and the self-similarity in groups of them can facilitate tackling of the inverse problem where an optimum pupil filter needs to be synthesized in accordance with prespecified diffraction characteristics.

References

1. Mukhopadhyay S, Sarkar S, Bhattacharya K, Hazra LN (2013) Polarization phase shifting interferometric technique for phase calibration of a reflective phase spatial light modulator. Opt Eng 52(3):035602-1–035602-6
2. Harmuth HF (1972) Transmission of information by orthogonal functions. Springer, Berlin, p 31
3. Beauchamp KG (1985) Walsh functions and their Applications. Academic Press, New York
4. Andrews HC (1970) Computer techniques in image processing. Academic, New York
5. Hazra LN, Banerjee A (1976) Application of Walsh function in generation of optimum apodizers. J Opt (India) 5:19–26
6. Hazra LN (1977) A new class of optimum amplitude filters. Opt Commun 21(2):232–236
7. De M, Hazra LN (1977) Real time image restoration through Walsh filtering. Opt Acta 24 (3):211–220
8. Hazra LN, Purkait PK, De M (1979) Apodization of aberrated pupils. Can J Phys 57(9):1340–1346
9. Hazra LN, Guha A (1986) Farfield diffraction properties of radial Walsh filters. J Opt Soc Am A 3(6):843–846
10. Hazra LN (2007) Walsh filters in tailoring of resolution in microscopic imaging. Micron 38 (2):129–135

Printed in the United States
By Bookmasters